Lecture Notes in Mathematics

Edited by A. Dold, Heidelberg and B. Eckmann, Zürich

396

Karl Heinrich H[illegible]
Michael Mislove
Albert Stralka

The Pontryagin Duality of Compact O-Dimensional Semilattices and its Applications

Springer-Verlag
Berlin · Heidelberg · New York 1974

Karl Heinrich Hofmann
Michael Mislove
Dept. of Mathematics
Tulane University
New Orleans, LA 70018/USA

Albert Stralka
Dept. of Mathematics
University of California
Riverside, CA 94202/USA

AMS Subject Classifications (1970): Primary: 06-02, 06A20, 20-02, 22A15
Secondary: 20M10

ISBN 3-540-06807-4 Springer-Verlag Berlin · Heidelberg · New York
ISBN 0-387-06807-4 Springer-Verlag New York · Heidelberg · Berlin

Offsetdruck: Julius Beltz, Hemsbach/Bergstr.

TO ALFRED HOBLITZELLE CLIFFORD
on his 65th birthday on the
11th of July, 1973

ACKNOWLEDGEMENTS

The authors were supported by NSF-Grants GP - 28655-A-1 and GP - 33912 during the preparation of this work. They are particularly grateful for the fact that this allowed Stralka to visit Tulane for a short period during which the general project of this research was outlined in a series of seminar sessions. The principal results were presented at the Houston conference on Lattice Theory in March, 1973. We also thank Mrs. Meredith R. Mickel for the excellent preparation of the typescript.

TABLE OF CONTENTS

INTRODUCTION

When Pontryagin established the duality between discrete and compact abelian groups in 1932 he was motivated by rather specific applications, mostly arising in an attempt at a general theory relating the following two examples from algebraic topology. Cech's homology groups of a compact space appeared as inverse limits of homology groups of finite complexes and thus behaved like compact abelian groups, whereas the discrete Cech cohomology groups arose from direct limits. The duality theory, however, evolved rather quickly to a rich structure theory which was applied to numerous areas of algebra, topology and analysis. In algebra and number theory these applications reach from Pontryagin's classification of the locally compact connected fields to the modern presentation of algebraic number theory (see W-1) while in group theory itself a rich interplay between the theory of abelian groups and compact groups developed giving impulses to both lines of research. Harmonic analysis, which had seen a great deal of activity during the twenties, was provided with precisely the right abstract tools by Pontryagin duality theory, and harmonic analysis became inseparable from the duality of locally compact abelian groups. Other dualities for various classes of topological groups followed, exemplified by the work of Tannaka and Krein in the thirties and forties, and the process of finding duality theories for general locally compact groups is still not completed.

In the theory of topological semigroups, which is less classical, duality theories have only lately been systematically investigated [H-4]. This is partly due to the fact that duality theories in the context of various classes of compact topological semigroups say, either do not exist, or, where indeed they do exist, are technically involved and rarely as simply expressed as Pontryagin duality [H-4]. However, in the case of semilattices, it was observed relatively early by Austin, as least on the object level, that an analog of Pontryagin character theory works [A-2], and this has been discussed in increasing measure by other authors (Baker and Rothman [B-1], Bowman [B-10], Hofmann [H-4], Schneperman [S-4]). In the meantime, duality theories for lattices and topological spaces in terms of characters were discussed by numerous authors in various degrees of generality; for a systematic treatment and further references in this direction see Hofmann and Keimel [H-5].

Nevertheless, it is pretty apparent, that the duality theory between discrete and compact semilattices, although having been treated from various angles, has never been systematically exploited. No applications to and from lattice theory have been made, and the theory of compact semigroups has not been brought to bear on this duality. The duality of semilattices should be a point where different lines of investigation merge: the algebraic theory of semilattices and lattices on one hand, and the theory of compact topological semigroups on the other.

In the following we make an attempt to present the duality theory of compact semilattices in this spirit. There are many features of this duality which place it in close parallel with Pontryagin duality; there are others in which it exhibits drawbacks, but also advantages.

We develop the general theory of the duality and we present applications to lattice theory on one hand and to compact monoid theory on the other. We hope that the future will bring applications beyond those which we know of and are able to discuss.

Various portions of the contents will be familiar to some groups of readers, varying along with the content; yet even what is likely to appear familiar is probably presented in a new light and, sometimes, in a more systematic fashion than any treatment other than that by duality would allow. What is perhaps well known to the person working in lattice theory may provide some new aspect for the worker in compact topological monoids, and vice versa. But even on familiar ground some new results emerge here and there.

The language of category theory provides a convenient and elegant medium for duality theory, and it places the emphasis where it belongs: on objects and functions equally. In the more conventional treatments of representations of semigroups or lattices, the role of the morphism is all to often ignored. However, the material entering from category theory is on the level of the theory of limits (touching upon the idea of Kan extensions of functors, although we will not explicitly speak about them) and the theory of adjoints. Very little of the deeper aspects of the theory of compact monoids will be needed, although the spirit of compact topological algebra pervades the discussion. Most of the lattice theory which appears is treated in a self-contained fashion, much of it with unconventional proofs.

The material is presented in the following fashion: In a preliminary Chapter, called Preliminaries, we provide the functorial language in which we will prove the duality. This involves functor categories, limit and colimit functors, and the concepts of density and codensity in categories. Beginning seriously with our topic, we open Chapter I with a section on the category of semilattices (with identity) and its basic properties; we then parallel this discussion in Section 2 by an analogous treatment of the category of compact zero dimensional semilattices (with identity). In Section 3 we prove the duality theorem; with our preparations the proof is very short and very apropos: We need to know that in both categories the finite objects determine the category in a sense which we make precise in terms of functorial density, and that the duality holds for finite semilattices. We do not need any intrinsic structural information about discrete or compact zero dimensional semilattices. This proof parallels a proof of the Pontryagin duality theorem for locally compact abelian groups which was recently given by Roeder [R-3]. Just as in the case of groups there are proofs of the duality theorem for semilattices available which utilize a considerable amount of structural information on semilattices. We prefer to present the proof which reflects the general set-up of numerous duality theorems, and then to go into the structural details for their own merit at a later time. In fact, this happens in Chapter II. In the first section we apply general principles from the theory of compact monoids to compact zero dimensional semilattices. In particular we record the monotone convergence theorem and the resulting long standing observation, that the underlying semilattice of a compact topological semilattice is a complete lattice. We introduce the concepts of a local minimum, a strong local minimum, and a semiminimum in a quasi-ordered set with a topology and their dual concepts. A local minimum is, as the name suggests, a minimum in one of its neighborhoods, a strong local minimum is an element which generates an open filter (upper set) and a semiminimum is a minimal element of the complement of some strong local maximum. In a topological semilattice, every local minimum is strong, but not every local maximum is strong. One of the important facts observed in this section is that the set $K(S)$ of local minima of a topological compact zero dimensional semilattice S is dense and closed under finite sups; in fact $s = \sup K(S) \cap Ss$ for each $s \in S$. It is also correct that every element is the inf of all semimaxima which dominate it. Thus there is an abundance of local minima and semimaxima. While there are plenty of local maxima, they do not play a role, and it may very well be the case that 1 is the only strong local maximum. On the other hand, there are also semilattices in which the set of local minima is dense but which nevertheless are not zero dimensional. In the second section of Chapter II we relate the concepts of characters and filters of a semilattice. In fact, beginning with a discrete semilattice, we set up an isomorphism between the compact zero dimensional character semilattice and the $\cap$-semilattice of all filters with a suitable topology. In this specific sense the character semilattice of a discrete semilattice is precisely the filter completion. As in all of our discussions we emphasize the functorial nature of such constructions. The third section of Chapter II is in many respects the core of the entire presentation. In the preceding section we described an alternative view of the characters of a discrete semilattice. Here we give an alternative description of a character of a compact zero dimensional discrete semilattice S; indeed there is an isomorphism between the character semilattice of S and the sup-semilattice K(S). The usefulness of this result is enhanced by a characterization theorem for the elements of K(S). Indeed an element k is a local minimum iff it is a compact element of the underlying complete lattice (whereby an element k of a lattice is compact if any relation $k \leq \sup X$ for a subset X of the lattice implies the existence of a finite subset $F \subseteq X$ with $k \leq \sup F$). Moreover this is equivalent to the statement that k is isolated in its principal ideal Sk and also to the statement that, in effect, it cannot be reached in Sk by any chain in $Sk \setminus \{k\}$. Recalling that a lattice is algebraic iff it is complete and every element is the sup of the compact elements which it dominates, we have now observed that the underlying lattice of any compact zero dimensional semilattice is an algebraic lattice. But we also demonstrate that, conversely, on every algebraic lattice there is a unique compact topology making it into a topological semilattice, and in this topology it is zero dimensional and the compact elements are precisely the local minima. Since we insist on the functorial aspect we have to find an algebraic description of the continuous morphisms: Indeed a semilattice morphism between two algebraic lattices is continuous relative to the unique topology just mentioned iff it is a lattice morphism preserving arbitrary infs and sups of chains. We call such morphisms algebraically continuous. Thus the principal results of this section and the Chapter may be summarized as follows: The category of compact zero dimensional semilattices (with identities) and continuous (identity preserving) semilattice morphisms is isomorphic to the category of algebraic lattices and algebraically continuous lattice maps and is dual to the category of all semilattices (with identities) and (identity preserving) semilattice morphisms. This result provides an algebraic basis for the entire theory. The question of a characterization of zero dimensional compact topological lattices (i.e. such objects in which the sup operation is also continuous) receives a partial answer in this section, too. This condition is satisfied if there is an abundance of strong maxima; unfortunately this condition is not purely algebraic, since there is no precise correspondence between the strong local maxima and the cocompact elements, and no equivalent formulation in terms of duality is known to us. However, as we will show in the later parts of the development, the situation becomes completely satisfactory in the presence of distributivity.

Chapters I and II contain the duality theory both in its general, i.e. category theoretical, and in its structural aspects. The remainder is devoted to applications. Chapter III links the theory with classical segments of lattice theory. The first Section of this Chapter is concerned with prime elements and distributivity; to no one's great surprise, these two concepts appear together. The concept of prime elements does not present any difficulty whatsoever in a semilattice, the concept of distributivity does. A semilattice has been called distributive if $\sup\{a,b\}x = \sup\{ax,bx\}$ whenever $\sup\{a,b\}$ exists. The nice aspect of this concept of distributivity is that it is characterized by the embeddability of the semilattice in a distributive lattice under preservation of existing sups. For the purpose of duality, however, this concept of distributivity is too weak, and we therefore call it the weak distributivity of a semilattice. Since distributivity involves sups we wish to bring the "virtual sups" into the play which exist in every semilattice (with identity), namely the filter $\uparrow a \cap \uparrow b$ where $\uparrow X$ denotes the filter of all elements dominating some element $x \in X$. We therefore define a semilattice to be distributive iff $\uparrow(\uparrow a \cap \uparrow b)x = \uparrow ax \cap \uparrow bx$ for all a,b,x in the semilattice. Every distributive semilattice is weakly distributive, the converse is false. One of the principal theorems of this section contains the result that a semilattice is distributive if and only if its (compact) character semilattice is distributive (and hence has a Brouwerian algebraic lattice as underlying semilattice). This is also equivalent to the property in the character semilattice that every element is the inf of the set of primes dominating it. On the other hand, a semilattice is primally generated in the sense that every element is a finite product (inf) of primes if and only if its character semilattice is a compact topological distributive lattice (i.e. has continuous sup-operation). This property of the character semilattice will be characterized in numerous other ways, the most purely algebraic being that its underlying lattice is bialgebraic, where we call a lattice bialgebraic if it is algebraic and if the opposite lattice (obtained by reversing the order) is also algebraic. The proofs of these facts are obtained through character theory. We define the concept of a sup-character (a special case of a sup-morphism between semilattices) which is a particular type of morphism preserving "virtual sups" (in the same spirit as we have used "virtual sups" to define distributivity). It turns out that a character of a semilattice is a sup-character iff it is a prime element of the character semilattice. This is the main link via duality between primes and distributivity, since distributivity implies the separation of points by sup-characters. However, one should be warned against assuming the converse; we are unable to prove it or furnish a counter example. Some of the principal results of the section may be summarized as follows: The category of distributive semilattices and morphisms mapping primes into primes is dual to the category of Brouwerian algebraic lattices and lattice morphisms preserving arbitrary sups and infs. The category of primally generated semilattices and prime preserving morphisms is dual to the category of Brouwerian bialgebraic lattices and lattice morphisms preserving arbitrary sups and infs (which category is isomorphic to the category of compact zero dimensional distributive topological lattices and continuous lattice morphisms). Some supplementary results link these facts with the categories of partially ordered sets and certain categories of topological spaces (the so-called spectral spaces). Section 2 sheds some light on the relation between the duality theory of semilattices and the classical theory of Boolean lattices. We prove that being Boolean and being free are dual properties in the following sense: The character semilattice of a free semilattice is a compact zero dimensional Boolean lattice with continuous multiplication (inf-operation); this implies in particular that every such object is of the form 2^X for some set X in the product topology. If, on the other hand, we start with a Boolean lattice and find the character semilattice of the underlying inf semilattice, then it turns out to be a free compact zero dimensional semilattice over a compact zero dimensional space. We say that a morphism between Boolean objects is Boolean iff it preserves complements. Since its dual is a morphism between free objects (in the sense explained) one naturally asks how these dual morphisms are characterized. In the case of free discrete semilattices, they are precisely the set induced morphisms; in the case of the free compact zero dimensional semilattices, they are precisely the space induced morphisms.

The third section of the chapter describes the projectives and injectives in the category of semilattices and its dual. These results complement the characterization theorems of Horn and Kimura, and the availability of duality enables us to give the proofs a different setting.

While Chapter III is focused on applications to algebra, the final Chapter IV illustrates various applications to topology, specifically the theory of compact topological semilattices and monoids. Since, in a sense, compact monoids were our point of departure in Chapter II when we began the structural investigation, this closes the circle. In Section 1 we use duality to determine "the topological size" of a compact zero dimensional semilattice. There are, among numerous others, two cardinals which are particularly useful to determine the size of a topology: one is the so-called weight, i.e. the smallest cardinal of a basis for the topology, and the other is what we call the separability number, i.e. the smallest cardinal of a dense subset. For a compact zero dimensional semilattice S we show that the weight $w(S)$ is the cardinal of its character semilattice $\hat{S}$, which by earlier results, equals card $K(S)$. If $d(S)$ is the separability number, we prove $\log w(S) \leq d(S) \leq w(S)$ (where $\log a = \min\{b \mid a \leq 2^b\}$); examples show that strict inequality occurs in both cases, but the estimates are the best possible. This contrasts the situation for compact abelian groups G: There one has $w(G) = \mathrm{card}(\hat{G})$ just as in the case of semilattices, but the equality $d(G) = \log w(G)$ is always true. Section 2 is a report on the application of duality to a very important line of research in the theory of compact monoids: the investigation of quotient morphisms which raise topological dimension. In fact, with the aid of duality, we are able to give a complete characterization of those compact zero dimensional semilattices which have quotients of positive topological dimension. They are precisely those which have the Cantor set chain (under min) as a quotient; dually, they are precisely those whose character semilattice contains an order dense non-degenerate countable chain. The results are in fact much sharper. The proofs of these results appear elsewhere, and we content ourselves here with a descriptive discussion of this theory. In the third and final section we prove another parallel theorem to one established for groups by various authors, in the most general form and with the most direct proof by Archangelski [A-1]. The theorem for groups says that a topological group whose underlying space is extremally disconnected is necessarily discrete. Here we show (by entirely different methods) that an extremally disconnected compact semilattice is necessarily finite. As a corollary of both results we then obtain the following theorem on compact semigroups: A compact monoid S which is a union of groups and is such that the set of idempotents is commutative cannot be extremally disconnected unless it is finite. As a biproduct of the results of this section we have an immediate proof of the fact that the space of closed subsets of an infinite compact extremally disconnected space is never extremally disconnected, since such a space is always a compact topological semilattice under $\cup$.

CHAPTER 0. Preliminaries

Section 1. About dense subcategories and the extension of natural transformations.

The background material which we are going to prepare here for later use may be presented in various degrees of generality. We choose that level which leads to our applications in the most direct fashion and does not require an apparatus of an exorbitant grade of abstraction. For a vastly more general approach to some of the ideas used here see [I-2].

We have to consider the category of diagrams in a given category. Let us denote with $\underline{cat}$ the category whose objects are small categories and whose morphisms are functors between them. The following lemmas have straightforward verifications (which can become a bit technical).

LEMMA 1.1. Let $\underline{A}$ be a category. The following definitions yield a category $\underline{A}^{\underline{cat}}$:

a) Objects: Objects are functors $D : \underline{X} \to \underline{A}$, $\underline{X} \in ob(\underline{cat})$

b) Morphisms: Morphisms $D \to E$, $D : \underline{X} \to \underline{A}$, $E : \underline{Y} \to \underline{A}$ are given by pairs $(f,a) : D \to E$ such that $f : \underline{X} \to \underline{Y}$ is a functor (i.e. a morphism in $\underline{cat}$) and $a : D \to Ef$ is a natural transformation of functors $\underline{X} \to \underline{A}$.

c) Composition: $(f,a)(g,b) = (fg,(ag)b)$

LEMMA 1.2. If $\phi : \underline{A} \to \underline{B}$ is a functor into a complete category, then there is an induced functor

$$\phi' = \phi^{\underline{cat}} : \underline{A}^{\underline{cat}} \to \underline{B}^{\underline{cat}}$$

$$\phi'(F) = \phi \circ F, \ \phi'(a,f) = (\phi a,f).$$

LEMMA 1.3. If $\underline{A}$ is cocomplete, then there is a functor $\mathrm{COLIM} : \underline{A}^{\underline{cat}} \to \underline{A}$ given by $\mathrm{COLIM}(D) = \mathrm{colim}\, D$ and $\mathrm{COLIM}(a,f) : \mathrm{colim}\, D \to \mathrm{colim}\, E$ given by

$$\begin{array}{ccccc} ((\mathrm{colim}\, D)f)_{\underline{Y}} & \longleftarrow & (\mathrm{colim}\, Df)_{\underline{Y}} & \xleftarrow{(\mathrm{colim}\, a)_{\underline{Y}}} & (\mathrm{colim}\, E)_{\underline{Y}} \\ \uparrow \lambda^{D} & & \uparrow \lambda^{Df} & & \uparrow \lambda^{E} \\ Df & = & Df & \xleftarrow[a]{} & E \end{array}$$

where for any object $A \in \underline{A}$, the constant functor $\underline{X} \to \underline{A}$ with value A is denoted by $A_{\underline{X}}$ and where $\lambda^{D} : D \to (\mathrm{colim}\, D)_{\underline{X}}$ denotes the limit natural transformation.

We now single out a subcategory of $\underline{A}^{\underline{cat}}$ which is of relevance in our context.

DEFINITION 1.4. Let $\underline{dir} \subseteq \underline{cat}$ be the full subcategory of all directed sets $\underline{X}$ made into a small category by writing $x \to y$ iff $x \le y$. Now $\underline{A}^{\underline{dir}}$ is the full subcategory in $\underline{A}^{\underline{cat}}$ of all direct systems, i.e. functors $D : \underline{X} \to \underline{A}$ with $\underline{X} \in \mathrm{ob}(\underline{dir})$. □

In topology (sheaf theory) and topological algebra, in fact, wherever limits more than colimits play a role, a variant of this set-up is more important; in duality theory we need both versions.

LEMMA 1.5. Let $(\underline{A}^{\underline{cat}})^{*} = [(\underline{A}^{op})^{\underline{cat}}]^{op}$. The objects of this category are functors $D : \underline{X} \to \underline{A}$ again, its morphisms are pairs $(f,a) : E \to D$, $E : \underline{Y} \to \underline{A}$, $D : \underline{X} \to \underline{A}$ with a functor $f : \underline{X} \to \underline{Y}$ and a natural transformation $a : Ef \to D$ of functors $\underline{X} \to \underline{A}$, which are composed according to $(g,b)(f,a) = (fg, b(ag))$.

If $\underline{A}$ is complete, then there is a functor $\mathrm{LIM} : (\underline{A}^{\underline{cat}})^{*} \to \underline{A}$ with $\mathrm{LIM}\,(D) = \lim D$ and with $\mathrm{LIM}\,(f,a)$ defined in a fashion analogous to $\mathrm{COLIM}\,(f,a)$ in 1.3.

This category contains a subcategory which will be of importance to us in a similar fashion as will be $\underline{A}^{\underline{dir}}$ in 1.4:

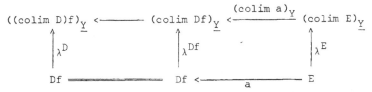

DEFINITION 1.6. Let $\underline{inv} \subseteq \underline{cat}$ be the full subcategory of all $\underline{X}^{op}$, $\underline{X} \in \underline{inv}$, i.e. the full subcategory of all inverse systems. Then $\underline{A}^{\underline{inv}}$ is the full subcategory in $(\underline{A}^{\underline{cat}})^*$.

Now we discuss briefly the concept of (directed) codensity and density.

DEFINITION 1.7. Let $\underline{A}_o$ be a subcategory of $\underline{A}$ with inclusion functor J. We say that $\underline{A}_o$ is codense in $\underline{A}$ (through direct limits) iff

i) $\underline{A}$ has direct limits (i.e. each $D \in \text{ob}\,\underline{A}^{\underline{dir}}$ has a colimit).

ii) there is a functor $\Delta : \underline{A} \to \underline{A}_o^{\underline{dir}}$ such that the functor

$$\underline{A} \xrightarrow{\ \Delta\ } \underline{A}_o^{\underline{dir}} \xrightarrow{\ J^{\underline{dir}}\ } \underline{A}^{\underline{dir}} \xrightarrow{\ \text{COLIM}\ } \underline{A}$$

is naturally isomorphic to the identity functor.

On the other hand, we say that $\underline{A}_o$ is dense in $\underline{A}$ (through projective limits) iff

i) $\underline{A}$ has projective limits (i.e. each $D \in \text{ob}\,\underline{A}^{\underline{inv}}$ has a limit).

ii) there is a functor $\nabla : \underline{A} \to \underline{A}_o^{\underline{inv}}$ such that the functor

$$\underline{A} \xrightarrow{\ \nabla\ } \underline{A}_o^{\underline{inv}} \xrightarrow{\ J^{\underline{inv}}\ } \underline{A}^{\underline{inv}} \xrightarrow{\ \text{LIM}\ } \underline{A}$$

is naturally isomorphic to the identity functor.□

We will encounter both of these situations in the context of our duality theory. The following is the relevant result we are using:

PROPOSITION 1.8. Let $\underline{A}_o$ be a codense [dense] subcategory of $\underline{A}$ (with inclusion functor J) and let $F,G : \underline{A} \to \underline{B}$ be two cocontinuous [continuous] (i.e. direct limit [projective limit] preserving) functors into a category with direct (projective) limits. Then for every natural transformation $\alpha_o : FJ \to GJ$ there is a unique natural transformation $\alpha : F \to G$ with $\alpha J = \alpha_o$ (i.e. α_o has a unique extension α).

Proof. Observe $F \cong F\ \mathrm{COLIM}\ J^{dir}\ \Delta \cong \mathrm{COLIM}(FJ)^{dir}\ \Delta$ and define an isomorphism $\mathrm{COLIM}\ \alpha_o{}^{dir}\ \Delta : \mathrm{COLIM}(FJ)^{dir}\ \Delta \to \mathrm{COLIM}\ (GJ)^{dir}\ \Delta$. For details see [H-3].

COROLLARY 1.9. Under the hypotheses of 1.8 if $\beta : F \to G$ is a natural transformation such that βJ is a natural isomorphism, then β itself is an isomorphism. In simplified terms: if the restrictions of two cocontinuous [continuous] functors to a codense [dense] subcategory are isomorphic, then they are isomorphic.

Let us note in passing that the theory outlined is a special instance of the theory of right and left Kan extensions of functors. This theory generally also provides the existence of extensions of functors, but we will not need this aspect here.

CHAPTER I. The category of discrete semilattices and the category of compact zero-dimensional semilattices

Section 1. The category of discrete semilattices.

In this section we introduce and discuss the category of semilattices with identity, but for completeness sake, we make the following definitions:

DEFINITION 1.1. A protosemilattice S is a commutative idempotent semigroup. For $s,t \in S$ we write $s \leq t$ if and only if $st = s$; this defines a partial order on S. A morphism of protosemilattices is a function $f : S \to T$ between protosemilattices satisfying $f(xy) = f(x)f(y)$. The class of all protosemilattices and their morphisms forms a category $\underline{PS}$. A semilattice is a protosemilattice with identity, and a morphism of semilattices is a protosemilattice morphism $f : S \to T$ between semilattices satisfying $f(1) = 1$. The category of all semilattices and semilattice morphisms will be denoted $\underline{S}$.

A subset S of a semilattice T is a subsemilattice if $SS \subseteq S$ and $1 \in S$ (if only $SS \subseteq S$, then S is called a proto-subsemilattice of T). If X is a subset of a protosemilattice S and if X has a greatest lower bound in S, then it will be denoted $\wedge X$ or $\inf X$.

We shall deal with protosemilattices rarely enough; however, we observe that this is no great loss of generality since one passes easily from each of the categories $\underline{S}$ and $\underline{PS}$ to the other:

PROPOSITION 1.2. The inclusion functor $\underline{S} \to \underline{PS}$ has as left adjoint the functor $\underline{PS} \to \underline{S}$ given by $S \mapsto S \cup 1$, where $S \cup 1$ is the semilattice obtained by adjoining an identity to S (i.e. by adjoining a new element $1 \notin S$ which operates on S by $1s = s1 = s$ for each $s \in S$); note that an identity is being adjoined regardless of the existence of an identity in S. Both functors are faithful, but neither is full, although the latter is representative (i.e. for each $S \in \underline{S}$ there is some $T \in \underline{PS}$ with $T \cup 1 \simeq S$).

Proof. The straightforward details are left as an exercise. □

NOTATION 1.3. The one, respectively, two element semilattice will be denoted 1, respectively, 2.

PROPOSITION 1.4. The category $\underline{S}$ is a pointed category with 1 as null object (i.e. as initial and coinitial object). The semilattice 2 is a generator and cogenerator for $\underline{S}$.

Proof. For $S \in \underline{S}$, the unique morphism $f \in \underline{S}(1,S)$ (respectively, $g \in \underline{S}(S,1)$) is given by $f(1) = 1$ (respectively, $g(s) = 1$ for each $s \in S$); hence 1 is the (essentially unique) null object. Let $f,g \in \underline{S}(S,T)$ with $f \neq g$. Then there is some $s \in S$ with $f(s) \neq g(s)$ and $s \neq 1$. Define $h \in \underline{S}(2,S)$ by $h(1) = 1$ and $h(0) = s$, and note that $fh \neq gh$, whence 2 is a generator for $\underline{S}$. Similarly, $f(s) \neq g(s)$ implies without loss of generality that $g(s) \not\geq f(s)$. Define $h \in \underline{S}(T,2)$ by $h(t) = \begin{cases} 1 & \text{if } t \geq f(s) \\ 0 & \text{otherwise} \end{cases}$. Then $hf \neq hg$, so 2 is also a cogenerator for $\underline{S}$. □

THEOREM 1.5. $\underline{S}(2,-) : \underline{S} \to \underline{Set}$ is a grounding functor from $\underline{S}$ to the category of sets, and it has a left adjoint $F : \underline{Set} \to \underline{S}$ given by $F(X) = (Fin(X), \cup)$ the semilattice of finite subsets of X under union with ϕ as identity; equivalently, $F(X) = {}^{X}2$ (the coproduct of X copies of 2). The front adjunction $X \to \underline{S}(2,F(X))$ is given by $x \mapsto f_x$ where $f_x(1) = \emptyset$ and $f_x(0) = \{x\}$; the back adjunction $F(\underline{X}(2,S)) \to S$ is given by $X \mapsto \wedge\{f(0): f \in X\}$.

Proof. First note that $\underline{S}(2,S)$ is isomorphic in $\underline{Set}$ to $|S|$, the underlying set of S under the function $f \mapsto f(0)$. We shall therefore denote $\underline{S}(2,S)$ by $|S|$ for each $S \in \underline{S}$. That $\underline{S}(2,-)$ is faithful is now clear.

For $X \in \underline{Set}$, the coproduct ${}^{X}2$ is just $\{f \in 2^X : \{x \in X : f(x) \neq 1\}$ is finite$\}$, and we define $\phi : F(X) \to {}^{X}2$ by $\phi(A) = \chi_{X \setminus A}$, the characteristic function of the set $X\setminus A$, which is in ${}^{X}2$ since A is finite. For $A,B \in F(X)$,

$$\phi(A \cup B) = \chi_{X\setminus(A\cup B)} = \chi_{(X\setminus A)\cap(X\setminus B)}$$
$$= \chi_{X\setminus A}\,\chi_{X\setminus B}$$
$$= \phi(A)\,\phi(B),$$ whence ϕ is

a homomorphism. Clearly the correspondence given by ϕ is a bijection.

Let $X \in \underline{\text{Set}}$ and $\eta : X \to |F(X)|$ by $\eta(x) = \{x\}$. Let $S \in \underline{S}$ and suppose $f \in \underline{\text{Set}}\,(X,|S|)$ is given. Define $f' \in \underline{S}(F(X),S)$ by $f'(A) = \wedge\{f(a) : a \in A\}$ for each $A \in F(X)$. Note that f' is well-defined since $A \in F(X)$ implies A is finite. For $A,B \in F(X)$,

$$f'(A \cup B) = \wedge\{f(x) : x \in A \cup B\}$$
$$= \wedge(\{f(x) : x \in A\} \cup \{f(x):x \in B\})$$
$$= (\wedge\{f(x) : x \in A\})\wedge(\wedge\{f(x):x \in B\})$$
$$= f'(A)\, f'(B),$$ whence

f' is an $\underline{S}$-morphism. Finally, for $x \in X$, $|f'|(\eta(x) = f'(\{x\}) = f(x)$, so $f = |f'|\eta$. Thus η is the front adjunction.

Similar straightforward calculation shows that $\varepsilon : F(|S|) \to S$ by $\varepsilon(A) = \wedge A$ is the back adjunction. □

PROPOSITION 1.6. A morphism f is $\underline{S}$ is monic if and only if $|f|$ is injective and epic if and only if $|f|$ is surjective.

Proof. Clearly $|f|$ injective implies f is monic, and since adjoint functors preserve monics, the converse is also true.

Suppose now that $f \in \underline{S}(S,T)$ for $S,T \in \underline{S}$ and let $t_o \in T\setminus f(S)$. Then $t_o < 1$ since $f(1) = 1$. Let $\chi_i \in \underline{S}(T,2)$, $i = 1,2$, by $\chi_1^{-1}(1) = \{t \in T : t_o \le t\}$ and $\chi_2^{-1}(1) = \uparrow\{t \in f(s) : t > t_o\}$. Then, for $s \in S$,

$$\chi_1 f(s) = \begin{cases} 1 & \text{if } t_o \le f(s) \\ 0 & \text{otherwise} \end{cases}$$
$$= \begin{cases} 1 & \text{if } t_o < f(s), \text{ since } t_o \notin f(s) \\ 0 & \text{otherwise} \end{cases}$$
$$= \chi_2 f(s),$$

but $\chi_1 \neq \chi_2$. Thus $|f|$ not surjective implies f is not epic, and since the converse always holds, we have the desired result. □

In combination with 1.6, the following theorem shows that $\underline{S}$ is just about as nice a category as one could desire.

THEOREM 1.7. (a). $\underline{S}$ is complete and cocomplete.
(b). The category $\underline{S}$ has biproducts; i.e. the canonical map $S \amalg T \to S \times T$ (existing in any pointed category) is an isomorphism.
(c). The hom sets $\underline{S}(A,B)$ are semilattices under the induced semilattice structure of $B^{|A|}$ (i.e. under pointwise operations).
(d). There is a tensor product of semilattices $(S,T) \mapsto S \otimes T : \underline{S} \times \underline{S} \to \underline{S}$, relative to which $(\underline{S}, \otimes)$ is a symmetric monoidal Cartesian closed category (i.e. $\otimes$ is commutative, associative, has 1 as identity object and satisfies the natural isomorphism $\underline{S}(A \otimes B,C) \cong \underline{S}(A,\underline{S}(B,C))$).

Proof. (a) The product of a family $\{S_j : j \in J\} \subseteq \underline{S}$ is just the Cartesian product under coordinate wise operations. If $f,g \in \underline{S}(S,T)$, then $E = \{s \in S : f(s) = g(s)\} \neq \emptyset$ as $1 \in E$. E is clearly the equalizer of f and g. Thus $\underline{S}$ is closed under arbitrary products and equalizers, so $\underline{S}$ is complete.

The coproduct of a family $\{S_j : j \in J\} \subseteq \underline{S}$, as remarked in the proof of 1.5, is just $\{(s_j) \in \prod_J S_j : s_j = 1$ for all but finitely many $j \in J\}$, and the coproduct map $c_i : S_i \to \coprod_J S_i$ is given by $(c_i(s))_j = \begin{cases} 1 & \text{if } i \neq j \\ s & \text{if } i = j \end{cases}$ for each $s \in S_i$. If $f,g \in \underline{S}(S,T)$, then the coequalizer of f and g is T/R where R is the smallest congruence on T containing $\{(f(s),g(s) : s \in S\}$. Therefore, since $\underline{S}$ has arbitrary coproducts and coequalizers, $\underline{S}$ is cocomplete.

(b) is clear in light of the description of the product and coproduct structures given in (a).

(c) is clear.

(d): For $X \in \underline{S}$, let $x \mapsto x' : X \to |F(X)|$ denote the front adjunction of 1.5. For $S,T \in \underline{S}$, consider on

$F(|S| \times |T|)$ the smallest congruence R identifying $(s_1s_2,t)'$ with $(s_1,t)'(s_2,t)'$ and $(s,t_1t_2)'$ with $(s,t_1)'(s,t_2)'$; equivalently, $R = \cap\mathcal{R}$ where $\mathcal{R}$ is the following family of congruences on $F(|S| \times |T|)$. Since $|S| \times |T| \subseteq F(|S| \times |T|)$ by the front adjunction, for a semilattice $L \in \underline{S}$, each $f \in \underline{\text{Set}}\ (|S| \times |T|, |L|)$ induces a homomorphism $f' \in \underline{S}\ (F(|S| \times |T|, |L|)$, and hence a congruence R_f on $F(|S| \times |T|)$, by $f'(A) = \wedge f(A)$ for each $A \in F(|S| \times |T|)$. Let $\mathcal{R}$ be the collection of all congruences R_f on $F(|S| \times |T|)$ induced by functions $f \in \underline{\text{Set}}$ with $f(s,t_1t_2) = f(s_1t_1)f(s_1t_2)$ and $f(s_1s_2,t) = f(s_1,t)f(s_2,t)$ for each $s,s_1,s_2 \in S$ and $t,t_1,t_2 \in T$. If we set $S \otimes T = F(|S| \times |T|)/R$, clearly then $S \otimes T$ has the universal property for bilinear maps: if $b \in \underline{S}(S \times T,L)$ is bilinear, then there is a unique $b' \in \underline{S}(S \otimes T, L)$ (which is described above) with $b'(s \otimes t) = b(s,t)$, where $s \otimes t = R((s,t)')$. The associativity, commutativity, and $S \otimes 1 \cong S \cong 1 \otimes S$, together with Mac Lane's coherence conditions follow as in the case of abelian groups. The isomorphism $\phi : \underline{S}(S \otimes T,U) \to \underline{S}(S,\underline{S}(T,U))$ is given by $\phi(f)(s)(t) = f(s \otimes t)$ with an inverse given by $\phi^{-1}(F)(s \otimes t) = F(s)(t)$. □

REMARK 1.8. Recall that a category is balanced if each monomorphism which is also an epimorphism is an isomorphism. Moreover, $\underline{S}(S,T)$ has an abelian semigroup structure by 1.7 (c). If $S,T,U \in \underline{S}$, then the composition functions $\underline{S}(T,U) \times \underline{S}(S,T) \to \underline{S}(S,U)$ are bilinear with respect to the above semigroup structure on $\underline{S}(S,T)$ and $\underline{S}(T,U)$, and the elements $1 \in \underline{S}(S,T)$ and $1 \in \underline{S}(T,U)$ given by $1(s) = 1$ and $1(t) = 1$ for each $s \in S$ and $t \in T$ act as zero morphisms with respect to composition. In the terminology of [M-4], $\underline{S}$ is then called semiadditive. A semiadditive category $\mathcal{A}$ in which $\mathcal{A}(A,B)$ is an abelian group for each $A,B \in \mathcal{A}$ is called an additive category. From the definitions and 1.6 it is clear that $\underline{S}$ is a balanced semiadditive category which is not additive. This is a counterexample to the last assertion of Proposition 18.4, p.30 of [M-5]. □

The category of semilattices has two particularly relevant "neighbor" categories which we now discuss. First, there is the category $\underline{PO}$ of partially ordered sets with maximum element and order preserving, maximum element preserving maps, and second, the category $\underline{DL}$ of distributive lattices with identity and identity preserving lattice morphisms.

PROPOSITION 1.9. The forgetful functors $\underline{DL} \xrightarrow{|\,|} \underline{S} \xrightarrow{|\,|} \underline{PO}$ have left adjoints $\underline{PO} \xrightarrow{\Sigma} \underline{S} \xrightarrow{\Lambda} \underline{DL}$ which are given as follows:

1.) For $(X,\le) \in \underline{PO}$, let $\Sigma(X,\le)$ be the set of all non-empty finite subsets of incomparable elements of X; i.e., $F \in \Sigma(X,\le)$ if F is a finite subset of X and $a,b \in F$, $a \le b$ implies $a = b$. For $F,G \in \Sigma(X,\le)$, define $FG = \min(F \cup G)$, the set of minimal elements of $F \cup G$. If $f : X \to Y$ is a morphism, then $(\Sigma f)(F) = \min f(F)$ for $F \in \Sigma(X,\le)$.
2.) First note that, for $L \in \underline{DL}$, $|L|$ is the semilattice L under its meet operation. Now, for $S \in \underline{S}$, let $\Lambda(S)$ be the semilattice of finitely generated ideals of S under $\cap$ and $\cup$. If $f : S \to T$ is a morphism and $I \in \Lambda(S)$, then $\Lambda(f)(I) = Tf(I)$.

Proof. 1.) For $(X,\le) \in \underline{PO}$, it is routine to check that $\Sigma(X,\le)$ is a semilattice with $\{1\}$ as identity, where 1 is the maximum of X. We define $\eta : (X,\le) \to |\Sigma(X,\le)|$ by $\eta(x) = \{x\}$. Suppose $S \in \underline{S}$ and $f \in \underline{PO}((X,\le), |S|)$. Define $f' \in \underline{S}(\Sigma(X,\le), S)$ by $f'(A) = \wedge f(A)$ for each $A \in \Sigma(X,\le)$. Then, for $A,B \in \Sigma(X,\le)$, $f'(AB) = \wedge f(AB) \le (\wedge f(A))(\wedge f(B)) = f'(A)\, f'(B)$. But, by definition of AB, if $x \in A \cup B$, there is $y \in AB$ with $y \le x$, so $f(y) \le f(x)$ as $f \in \underline{PO}$. Thus, $f'(AB) = \wedge f(AB) \le f(y) \le f(x)$. As $x \in A \cup B$ is arbitrary, $f'(AB) \le \wedge f(A \cup B) = f'(A)f'(B)$. Therefore $f'(AB) = f'(A)f'(B)$, whence $f' \in \underline{S}$. Finally, for $x \in X$, $|f'|\eta(x) = |f'|(\{x\}) = f(x)$, so $|f'|\eta = f$.

Similar routine verification shows that the back adjunction $\varepsilon : \Sigma(|S|) \to S$ is given by $\varepsilon(A) = \wedge A$ for each $A \in \Sigma(|S|)$.

2.) For $S \in \underline{S}$, clearly the collection $\Lambda(S)$ of finitely generated ideals of S is a distributive lattice under $\cap$ and $\cup$ with $S = S1$ as identity. Define $\eta \in \underline{S}(S, |\Lambda(S)|)$ by $\eta(s) = Ss$, the ideal generated by s. Since for $s_1, s_2 \in S$, $Ss_1 \cap Ss_2 = Ss_1s_2$, η is an $\underline{S}$-morphism. Suppose $L \in \underline{DL}$ and $f \in \underline{S}(S, |L|)$. In order to define $f' \in \underline{L}(\Lambda(S), L)$ with $f = |f'|\eta$, we make the following observation. Each $I \in \Lambda(S)$ is of the form $I = Ss_1 \cup Ss_2 \cup \ldots \cup Ss_n$ for some collection $\{s_1, s_2, \ldots, s_n\} \subseteq S$. Moreover, if $Ss_1 \cup \ldots \cup Ss_n = St_1 \cup \ldots \cup St_m$ for some collection $\{t_1, \ldots, t_m\}$, then $\{t_1, \ldots, t_m\} \subseteq Ss_1 \cup \ldots \cup Ss_n$ and $\{s_1, \ldots, s_n\} \subseteq St_1 \cup \ldots \cup St_m$. Thus, for each $i \in \{1, \ldots, n\}$, there is some $j \in \{1, \ldots, m\}$ with $s_i \in St_j$, and for each $j \in \{1, \ldots, m\}$ there is some $i \in \{1, \ldots, n\}$ with $t_j \in Ss_i$. Now, define $f' : \Lambda(S) \to L$ by $f'(Ss_1 \cup \ldots \cup Ss_n) = \vee\{f(s_1), \ldots, f(s_n)\}$. If $Ss_1 \cup \ldots \cup Ss_n = St_1 \cup \ldots \cup St_m$, then, by the above, for each $j \in \{1, \ldots, m\}$, there is some $i \in \{1, \ldots, n\}$ with $t_j \in Ss_i$, whence $f(t_j) \in Lf(s_i)$ as f is an $\underline{S}$-morphism. Thus, $f(t_j) \leq \vee\{f(s_i) : i = 1, \ldots, n\}$, and since $j \in \{1, \ldots, m\}$ is arbitrary, $\vee\{f(t_j) : j\ 1, \ldots, m\} \leq \vee\{f(s_i) : i = 1, \ldots, n\}$. The reverse inequality follows similarly, and so f' is well-defined.

Suppose that $Ss_1 \cup \ldots \cup Ss_n, St_1 \cup \ldots \cup St_m \in \Lambda(S)$. Then,
$f'[(Ss_1 \cup \ldots \cup Ss_n) \cap (St_1 \cup \ldots \cup St_m)]$

$$= f'[\bigcup_{i,j} (Ss_i \cap St_j)] = f'[\bigcup_{i,j} (Ss_it_j)]$$

$$= \vee\{f(s_it_j) : i = 1, \ldots, n;\ j = 1, \ldots, m\}$$

$$= \vee\{f(s_i) \wedge f(t_j) : i = 1, \ldots, n;\ j = 1, \ldots, m\}$$

$$= (\vee\{f(s_i) : i = 1, \ldots, n\}) \wedge (\vee\{f(t_j) : j = 1, \ldots, m\}),$$

since L is distributive

$$= f'(Ss_1 \cup \ldots \cup Ss_n) \wedge f'(St_1 \cup \ldots \cup St_m).$$

Thus f' is a $\underline{DL}$-morphism. Finally, for $s \in S$, $|f'|\eta(s) = |f'|(Ss) = f(s)$, so $f = |f'|\eta$.

Again, similar routine calculations show that, for $L \in \underline{DL}$, the map $\varepsilon : \Lambda(|L|) \to L$ given by

$\varepsilon(|L|\wedge\ell_1 \cup \ldots \cup |L|\wedge\ell_n) = \vee\{\ell_i : i = 1,\ldots,n\}$ is the back adjunction. □

REMARK 1.10. On the surface, it might seem more natural to set up an adjunction between $\underline{S}$ and $\underline{L}$, the category of lattices with identity and identity preserving lattice morphisms. Although such an adjunction does exist (see Exercise 1.14), we are unable to find a concrete realization of the adjoint to the grounding functor $\underline{L} \xrightarrow{||} \underline{S}$. □

We now establish a key lemma in the duality theory we are pursuing.

DEFINITION 1.11. Let fin $\underline{S}$ denote the full subcategory of finite semilattices in $\underline{S}$.

PROPOSITION 1.12. The category fin $\underline{S}$ is codense in the category $\underline{S}$. Specifically, let $\Delta : \underline{S} \to (\text{fin } \underline{S})^{\underline{\text{dir}}}$ be the functor which associates with a semilattice $S \in \underline{S}$ the direct system $\Delta(S)$ of all finite subsemilattices of S under inclusion, considered as a diagram $\Delta(S) : \text{Sub}(S) \to \text{fin } \underline{S}$, where Sub(S) denotes the directed system of finite subsemilattices of S. If $f \in \underline{S}(S,T)$, then $\Delta(f) = (\bar{f},\eta)$ where $\bar{f} : \text{Sub}(S) \to \text{Sub}(T)$ is given by $\bar{f}(F) = f(F)$ for each $F \in \text{Sub}(S)$, and where $\eta_F : \Delta(S)(F) \to \Delta(T)\bar{f}(F)$ is the natural surjection $\eta_F = f|_F : F \to f(F)$. Then $\text{COLIM } J\Delta = I$, where $J : \text{fin } \underline{S} \to \underline{S}$ is the inclusion and I is the identity.

Proof. Straightforward. □

EXERCISES

EXERCISE 1.13. Let $\underline{S}^1 = \{f \in \underline{S} : f^{-1}(1) = \{1\}\}$. There are functors $\underline{S}^1 \to \underline{PS}$ and $\underline{PS} \to \underline{S}^1$ given by $S \in \underline{S}^1 \longmapsto S\setminus\{1\} \in \underline{PS}$ and $S \in \underline{PS} \longmapsto S \cup 1 \in \underline{S}^1$. These functors form an equivalence between $\underline{S}^1$ and $\underline{PS}$, indeed the composition of the functors in either order is the identity on the appropriate category.

EXERCISE 1.14. Let $\underline{L}$ be the category of lattices with (meet) identity and identity preserving lattice morphisms, and let $\underline{L} \xrightarrow{||} \underline{S}$ be the grounding functor which associates to a lattice $L \in \underline{L}$ its meet semilattice. Then there is a left adjoint $\Lambda : \underline{S} \to \underline{L}$ to $|\,| : \underline{L} \to \underline{S}$. (Use Freyd's left adjoint existence theorem.)

EXERCISE 1.15. The category $\underline{S}$ contains as a subcategory, the category $\underline{S}_o$ of all semilattices with 0 and 0-preserving $\underline{S}$-morphisms. The functor associating with a semilattice $S \in \underline{S}$ the semilattice $S \in \underline{S}_o$ obtained by adjoining a 0 to S (whether S has one or not) is a full and faithful left reflector of $\underline{S}$ into $\underline{S}_o$. A similar statement holds for $\underline{PS}$ and $\underline{PS}_o$.

Section 2. The category of compact zero-dimensional semilattices

We now introduce our proposed dual category for S and prove several results analogous to those established for $\underline{S}$ in Section 1.

DEFINITION 2.1. A compact zero dimensional (cz) protosemilattice S is a compact zero dimensional space together with a continuous semilattice multiplication $S \times S \to S$. A morphism of cz-protosemilattices is a continuous map $f : S \to T$ preserving the operation. Accordingly, a cz-semilattice is a cz-protosemilattice with an identity, a cz-semilattice morphism preserves the identity, as well as being continuous and operation preserving.

The category of cz-semilattices and cz-semilattice morphisms will be denoted with $\underline{Z}$; the larger category of cz-protosemilattices and cz-protosemilattice morphisms by $\underline{PZ}$.

PROPOSITION 2.2. The inclusion functor $\underline{Z} \to \underline{PZ}$ is faithful, but not full; it has a left adjoint $\underline{PZ} \to \underline{Z}$ given by $S \to S \cup 1$, where $S \cup 1$ denotes the result of adjoining an isolated identity to S (again, irrespective of whether or not S has one). The left adjoint is faithful, but not full.

Proof. Let $S \in \underline{Z}$, and define $\varepsilon : S \cup 1 \to S$ by $\varepsilon(s) = s$ if $s \in S$ and $\varepsilon(1) = 1_S$, where 1_S is the identity of S. Let $T \in \underline{PZ}$ and suppose $f \in \underline{Z}\,(T \cup 1, S)$. Then $f' = f \mid T : T \to S$ is a $\underline{PZ}$-morphism, and its image under the functor $\underline{PZ} \to \underline{Z}$ is the $\underline{Z}$-morphism $f'' : T \cup 1 \to S \cup 1$ given by $f''(t) = f(t)$ for $t \in T$ and $f''(1) = 1$. Thus, for $t \in T$, $\varepsilon f''(t) = \varepsilon(f(t)) = f(t)$ as $f(t) \in S$, and $\varepsilon f''(1) = 1_S$. This shows that ε is the back adjunction.

A similar calculation shows that, for $S \in \underline{PZ}$, the front adjunction $\eta : S \to S \cup 1$ is given by $\eta(s) = s$ for each $s \in S$.

That each functor is faithful but not full is left as a (trivial) exercise. □

DEFINITION 2.3. Let fin $\underline{Z}$ be the full subcategory of $\underline{Z}$ of finite semilattices.

Note that fin $\underline{Z}$ = fin $\underline{S}$.

The following proposition is crucial not only to our proof of the duality between $\underline{S}$ and $\underline{Z}$, but also to the proofs of most of the remaining results of this section.

PROPOSITION 2.4. The category fin $\underline{Z}$ is dense in $\underline{Z}$. Specifically, let $\nabla : \underline{Z} \to (\text{fin } \underline{Z})^{\underline{\text{inv}}}$ be the functor which associates the inverse system $\nabla(S)$ of all finite quotients S/R with a semilattice $S \in \underline{Z}$, this system considered as a diagram $\nabla(S) : \text{Cong}(S) \to \text{fin } \underline{Z}$ with $\nabla(S)(R) = S/R$, where $\text{Cong}(S)$ is the inverse system of congruences R on S with finitely many congruence classes. If $f \in \underline{Z}(S,T)$ for $S,T \in \underline{Z}$, then $\nabla(f) = (\bar{f}, \eta)$, where $\bar{f} : \text{Cong}(T) \to \text{Cong}(S)$ is the functor given by $\bar{f}(R) = (f \times f)^{-1}(R)$, and where $\eta_R : \nabla(S)\bar{f}(R) \to \nabla(T)(R)$ is the natural injection $S/\bar{f}(R) \to T/R$. Then $\text{LIM } J_{\underline{Z}} \nabla = I_{\underline{Z}}$, where $J_Z : \text{fin } \underline{Z} \to \underline{Z}$ is the inclusion.

Proof. It has been proved by Numakura for the first time [N-2] (see also [H-7]) that every compact zero-dimensional topological semigroup S is the projective limit of all of its finite quotients S/R. Therefore, it remains to verify that the functorial properties of the density situation are satisfied (0-1.7). This is done by straightforward checking (a similar example has been worked out in Hofmann [H-3], pp. 117, 118.)

We now present a series of results which parallel

those obtained for $\underline{S}$ in Section 1.

PROPOSITION 2.5. The category $\underline{Z}$ is a pointed category with 1 as null object. The object 2 is a generator and cogenerator for $\underline{Z}$.

Proof. The proofs of all statements except the cogeneration by 2 are the same as those given for $\underline{S}$ in Section 1. Since fin $\underline{Z}$ = fin $\underline{S}$, 2 is a cogenerator for fin $\underline{Z}$; it follows from 2.4 that 2 is also a cogenerator for $\underline{Z}$. □

PROPOSITION 2.6. Let $\underline{Z}$ $\underline{\text{Comp}}$ be the category of zero dimensional compact Hausdorff spaces, and for $X \in \underline{Z}$ $\underline{\text{Comp}}$, let $\Gamma(X)$ be the space of closed subsets of X in the exponential topology (see [K-2], pp. 160.) and with union as operation. Then $\Gamma(X) \in \underline{Z}_1$ and $\Gamma : \underline{Z}\ \underline{\text{Comp}} \to \underline{Z}$ is the left adjoint of the forgetful functor $|\,| : \underline{Z} \to \underline{Z}\ \underline{\text{Comp}}$. The front adjunction $\eta : X \to |\Gamma(X)|$ is given by $\eta(x) = \{x\}$, while the back adjunction $\varepsilon : \Gamma(|S|) \to S$ is given by $\varepsilon(A) = \wedge A$ for each $A \in \Gamma(|S|)$, for each $S \in \underline{Z}$.

Proof. We must verify the following universal property: If $S \in$ ob $\underline{Z}$, $X \in$ ob $\underline{Z}$ $\underline{\text{Comp}}$, and if $f\colon X \to |S|$ is a continuous function, then there is a unique $\underline{Z}$-morphism $f'\colon \Gamma(X) \to S$ with $f(x) = f'(\{x\})$. If $f'\colon \Gamma(X) \to S$ is given by $f'(A) = \wedge A$, then f' is defined as S is compact, and f' is a semilattice morphism. Then f' is the composition of the $\underline{Z}$-morphism $\Gamma(f) : \Gamma(X) \to \Gamma(|S|)$ and the semilattice morphism $\inf : \Gamma(|S|) \to S$. It remains to show the continuity of inf; then f' is continuous and then also unique relative to its properties. It is a general fact that for a compact semilattice S, the semilattice morphism $\inf : \Gamma(|S|) \to S$ is continuous if and only if S has small semilattices (i.e. every $s \in S$ has a neighborhood basis consisting of proto-subsemilattices). However, we will give a direct proof in the zero dimensional case: The functor Γ preserves surjectivity; it is easy to conclude from this observation that Γ also preserves strict projective limits (i.e. limits of inverse systems with surjective maps) (see e.g. [H-8], Lemma III in the proof of 1.8). Let $\lambda_j : S \to S_j$ be the surjective limit maps onto the objects of the strict projective system of all finite quotients of S. Then we have a commutative diagram

$$\begin{array}{ccc}
\Gamma(S) = \lim \Gamma(S_i) & \xrightarrow{\ \inf_S\ } & \lim S_j = S \\
{\scriptstyle \Gamma(\lambda_j)}\downarrow & & \downarrow{\scriptstyle \lambda_j} \\
\Gamma(S_j) & \xrightarrow[\ \inf_{S_j}\ ]{} & S_j
\end{array}$$

in which the inf operation appears as the unique limit fill-in map; therefore $\inf_S$ is a $\underline{Z}$-morphism. □

PROPOSITION 2.7. The grounding functor $\underline{Z} \to \underline{\text{Set}}$ is given by $\underline{Z}(2,-)$. It has a left adjoint $\Phi : \underline{\text{Set}} \to \underline{Z}$ which is given by $\Phi = \Gamma \circ \beta$, where $\beta : \underline{\text{Set}} \to \underline{\text{Z Comp}}$ is the Stone-Čech compactification functor.

Proof. Since left adjoints compose, the result follows from 2.6. □

PROPOSITION 2.8. A $\underline{Z}$-morphism f is monic if and only if $|f|$ is injective and epic if and only if $|f|$ is surjective.

Proof. Since $\underline{Z}$ has a free functor $\Phi : \underline{\text{Set}} \to \underline{Z}$, monics are precisely the injectives.

Clearly surjectives are epics. Conversely, suppose $S,T \in \underline{Z}$ and $f \in \underline{Z}(S,T)$ is epic. Assume that f is not surjective. Then, by 2.4, there is a congruence R on T with finitely many cosets and quotient map $q : T \to T/R$ such that qf is not surjective. But qf is epic, and if R' is the kernel congruence of qf, then $S/R' \to T/R$ is an epic, non-surjective morphism in fin $\underline{Z} \subseteq \underline{S}$, contradicting 1.6. □

<u>THEOREM 2.9</u>. a.) The category $\underline{Z}$ is complete and cocomplete

b.) $\underline{Z}$ has biproducts (see (1.7).

c.) The hom-sets $\underline{Z}(S,T)$ are semilattices under the induced semilattice structure of $T^{|S|}$.

d.) There is a tensor product of $\underline{Z}$-semilattices $(S,T) \to S \otimes T : Z \times \underline{Z} \to \underline{Z}$ relative to which $(\underline{Z}, \otimes)$ is a symmetric monoidal category. Moreover, for $S,T \in \underline{Z}$, $S \otimes T$ has the universal property that for each bilinear continuous map $f : S \times T \to U$ of $S \times T$ to a $\underline{Z}$-object U, there is a unique $\underline{Z}$-morphism $f' : S \otimes T \to U$ with $f(s,t) = f'(s \otimes t)$ for each $(s,t) \in S \times T$.

Proof. a.) The product in $\underline{Z}$ of a family of $\underline{Z}$-objects is the Cartesian product endowed with the Tychonoff topology and coordinate-wise multiplication, while the equalizer of two $\underline{Z}$-morphisms $f,g : S \to T$ is the set of points of S where they agree. As $\underline{Z}$ has products and equalizers, $\underline{Z}$ has limits, so is complete.

At a later stage, as soon as we know that $\underline{S}$ and $\underline{Z}$ are dual categories, we shall derive the cocompleteness of $\underline{Z}$ from the completeness of $\underline{S}$. In the meantime, let $D : I \to \underline{Z}$ be a diagram (functor with small domain). Since $\underline{Z\ Comp}$ is cocomplete, the diagram $|D| : I \to \underline{Z\ Comp}$ has a colimit, $\operatorname{colim} |D|$ with colimit maps $c_i : |D(i)| \to \operatorname{colim}|D|$. On the $\underline{Z}$-object $S = \Gamma(\operatorname{colim} |D|)$ we consider the smallest congruence R which identifies $\{c_i(x,y)\}$ with $\{c_i(x)\} \cup \{c_i(y)\}$ for $x,y \in D(i)$ for each i. The functions $c_i' = qc_i$, where $q : S \to S/R$ is the quotient morphism, are then $\underline{Z}$-morphisms, and it is easily verified that S/R is the colimit of D with colimit maps c_i'.
b.) follows from the fact that the coproduct of S and T is just $S \times T$ for $\underline{Z}$-objects S and T. c.) is clear, and the proof of d.) which is similar to the spirit of that of (1.7 d) and relies on 2.7 is left as an exercise. □

In closing, we note that $\underline{Z}$ is not Cartesian closed relative to $\otimes$, since $\underline{Z}(S,T)$ is not in $\underline{Z}$ in general, even if we equip the hom-set with one of the standard topologies such as the compact-open topology.

Section 3. Characters and the duality between $\underline{S}$ and $\underline{Z}$

In this section we introduce the notion of character in $\underline{S}$ and in $\underline{Z}$ and bring together our previous results to obtain the duality of $\underline{S}$ and $\underline{Z}$.

DEFINITION 3.1. If $S \in \underline{S}$, then a character of S is a morphism $c : S \to 2$, i.e. an element of $\underline{S}(S,2)$. Similarly, a character of $\underline{Z}$-semilattice S is a $\underline{Z}$-morphism $c : S \to 2$, i.e. an element of $\underline{Z}(S,2)$.

LEMMA 3.2. If S is an object of $\underline{S}$ ($\underline{Z}$), then the $\underline{S}$ (respectively, $\underline{Z}$) characters separate the points of S.

Proof. Let $S \in \underline{S}$ and $s,t \in S$ with $s \neq t$, and assume $st < t$. If $c : S \to 2$ is defined by $c^{-1}(1) = \uparrow t = \{x \in S : xt = t\}$, then c is a character which separates s and t. Thus the statement holds for those $S \in \underline{S}$. But $\text{fin}\ \underline{Z} = \text{fin}\ \underline{S} \subsetneq \underline{S}$, whence the characters separate the points for objects in $\text{fin}\ \underline{Z}$. Since $\text{fin}\ \underline{Z}$ is dense in $\underline{Z}$ by 2.4, the characters separate the points for each $S \in Z$.□

LEMMA 3.3. For each set X, the product 2^X is a $\underline{Z}$-object. If $S \in \underline{S}$, then $\underline{S}(S,2)$ is a closed subset of $2^{|S|}$, and is therefore a $\underline{Z}$-object.

Proof. Exercise. □

DEFINITION 3.4. For $S \in \underline{S}$, the semilattice $\underline{S}(S,2)$ with the (compact) topology of pointwise convergence will be called the character semilattice of S and denoted by $\hat{S}$. For $S \in \underline{Z}$, the $\underline{S}$-object $\underline{Z}(S,2)$ will be called the character semilattice of S and will also be denoted by $\hat{S}$.

LEMMA 3.5. If $f \in \underline{S}(S,T)$ for $\underline{S}$-objects S and T, we define $\hat{f} \in \underline{Z}(\hat{T},\hat{S})$ by $\hat{f}(c) = c \circ f$. Then $\hat{} : \underline{S} \to \underline{Z}^{op}$ is a functor. The same definitions apply with the roles of $\underline{S}$ and $\underline{Z}$ interchanged.

Proof. Again, we leave this as an easy exercise. □

LEMMA 3.6. For $S \in \underline{S}$ there is a natural injective transformation $\eta_S : S \to \hat{\hat{S}}$ given by $\eta_S(s)(c) = c(s)$, and for $S \in \underline{Z}$ there is a natural injective transformation $\varepsilon_S : S \to \hat{\hat{S}}$ also given by $\varepsilon_S(s)(c) = c(s)$.

Proof. For $S \in \underline{S}$, clearly $\eta_S : S \to \hat{\hat{S}}$ is an $\underline{S}$-morphism, and its injectivity follows from 3.2. Similarly, for $S \in \underline{Z}$ $\varepsilon_S : S \to \hat{\hat{S}}$ is an injective $\underline{S}$-morphism whose continuity is also clear. □

LEMMA 3.7. Let $S \in \text{fin}\ \underline{S} = \text{fin}\ \underline{Z}$. Then there is a bijection $s \mapsto f_s : S \to \hat{S}$ given by $f_s^{-1}(1) = \uparrow s = \{t \in S : s \leq t\}$. In particular, card S = card $\hat{S}$.

Proof. The correspondence $s \mapsto f_s$ for $s \in S$ is a well-defined function mapping S into $\hat{S}$. If $f_s = f_t$, then $\uparrow s = \uparrow t$ whence $s = t$. Thus the correspondence is also one to one. If $c \in \hat{S}$, then $c^{-1}(1)$ is a subsemilattice of S which has an infimum, s, as S is finite. Clearly, then, $c = f_s$, so the correspondence is also surjective. □

LEMMA 3.8. For $S \in$ fin $\underline{S}$ = fin $\underline{Z}$, $\eta_S = \varepsilon_S : S \to \hat{\hat{S}}$ is an isomorphism.

Proof. $\eta_S = \varepsilon_S$ is injective by 3.6, and card S = card $\hat{\hat{S}}$ by 3.7. □

THEOREM 3.9. (The Duality Theorem). The categories $\underline{S}$ and $\underline{Z}$ are dual under the functors $\hat{} : \underline{S} \to \underline{Z}^{op}$ and $\hat{} : \underline{Z} \to \underline{S}^{op}$, and the natural transformations η and ε are isomorphisms.

Proof. Let $J_{\underline{S}}$: fin $\underline{S} \to \underline{S}$ be the inclusion functor . The category fin $\underline{S}$ is codense in $\underline{S}$ by 1.12, and the natural transformation $\eta J_{\underline{S}} : J_{\underline{S}} \to \hat{}\hat{}J_{\underline{S}}$ is an isomorphism by 3.8. Hence η_S is an isomorphism by 0-1.9.

Likewise, the category fin $\underline{Z}$ is dense in $\underline{Z}$ by 2.4 and the natural transformation $\eta J_{\underline{Z}} : J_{\underline{Z}} \to \hat{}\hat{}J_{\underline{Z}}$ is an isomorphism by 3.8. Hence ε_S is an isomorphism by 0-1.9. □

We close this section with another adjunction between $\underline{S}$ and $\underline{Z}$ which utilizes the obvious grounding functor from $\underline{Z}$ to $\underline{S}$.

THEOREM 3.10. The functor $U : \underline{Z} \to \underline{S}$ assigning to each $S \in \underline{Z}$ the underlying semilattice $U(S) \in \underline{S}$ has a left adjoint $\alpha : \underline{S} \to \underline{Z}$ given as follows: For $S \in \underline{S}$, $\alpha(S) = U(\hat{S})\hat{}$. The front adjunction $\gamma : S \to U(U(\hat{S})\hat{})$ given by $\gamma(s)(c) = c(s)$ is a natural injection.

Proof. Let $S \in \underline{S}$, $T \in \underline{Z}$ and $f \in \underline{S}(S, U(T))$. Define a unique $\underline{S}$-morphism $\tilde{f} : \hat{T} \to U(\hat{S})$ by $\tilde{f}(c)(s) = c(f(s))$. This in turn yields a unique $f' : U(\hat{S})\hat{} \to T$ so that

$$\begin{array}{ccc} U(\hat{S})^{\wedge} & \xrightarrow{\ f'\ } & T \\ & {\scriptstyle \tilde{f}^{\wedge}} \searrow \quad \swarrow {\scriptstyle \varepsilon_T} & \\ & \hat{\hat{T}} & \end{array}$$

commutes. (Recall that ε_T is an isomorphism.) Now, for $s \in S$, $[\varepsilon_T f'(s)]\,(c) = c(f(s))$ for $c \in \hat{T}$, while, on the other hand, $[\tilde{f}^{\wedge} r(s)](c) = [r(s) \circ \tilde{f}](c) = r(s)(f(c)) = f(c)(s) = c(f(s))$. This shows r is indeed the front adjunction. □

DEFINITION 3.11. α is called the Bohr compactification functor, and $\alpha(S)$ the Bohr compactification of S.

EXERCISES

EXERCISE 3.13. A bialgebra over the field of complex numbers $\mathbb{C}$ is an algebra A over $\mathbb{C}$ with coassociative algebra map $c : A \to A \otimes A$. If c is compatible with the switch automorphism of $A \otimes A$, then c is called cocommutative. A coidentity is an augmentation $u : A \to \mathbb{C}$ such that $(A \otimes u)c : A \to A \otimes \mathbb{C}$ is the natural injection given by $a \mapsto a \otimes c$ (and similarly on the other side). A bialgebra is called idempotent iff mc is the identity map of A where $m : A \otimes A \to A$ represents the multiplication. Bialgebra morphisms are defined in a natural fashion.

The category S of semilattices is equivalent to the category of biregular commutative, cocommutative, idempotent $\mathbb{C}$-algebras with identity and coidentity. (A commutative algebra is biregular iff it is regular iff every principal ideal is generated by an idempotent.) For this result and further background, see [H-4].

The principal features of the duality theorem carry over to wider categories.

EXERCISE 3.13. If S is a locally compact zero dimensional semilattice, then the continuous characters $S \to 2$ separate the points of S. [B-1].

EXERCISE 3.14. On the semilattice $\hat{S}$ of continuous characters on S we consider the compact open topology. This topology has a subbasis consisting of sets of the type

$\{f \in S \mid f(K) = \{0\}\}$, $K \subseteq S$ compact and of the type $\{f \in S \mid f(s) = 1\}$, $s \in S$. [B-10].

For more details on locally compact zero dimensional semilattices see Schneperman, L. B., [S-4].

Section 4. General consequences of duality

In this section we collect some more or less immediate consequences of the duality and of the basic facts described earlier in this Chapter.

PROPOSITION 4.1. Let $f : S \to T$ be in $\underline{S}$ and $g = \hat{f} : \hat{T} \to \hat{S}$ its dual in $\underline{Z}$. Then

(a) f is injective iff g is surjective.

(b) f is surjective iff g is injective.

(c) f is a retraction (i.e. has a right inverse) iff g is a coretraction (i.e. has a left inverse).

(d) f is an isomorphism iff g is an isomorphism.

Proof. In view of 1.6 and 2.8, all of these assertions are immediate.

PROPOSITION 4.2. Let $D : I \to \underline{S}$ be a diagram (a functor) and let $\hat{D} : I \to \underline{Z}$ be the dual diagram $\hat{} \circ D$. Then we have the following conclusions:

(a) $S = \lim D$ with $\lambda : (\lim D)_I \to D$ the limit natural transformation (where $S_I : I \to \underline{S}$ denotes the constant functor with value S) iff $\hat{S} = \operatorname{colim} \hat{D}$ with $\hat{\lambda} : \hat{D} \to \hat{S}$ the colimit natural transformation.

(b) $S = \operatorname{colim} D$ with $\kappa : D \to (\operatorname{colim} D)_I$ the colimit map iff $\hat{S} = \lim \hat{D}$ with $\hat{\kappa} : \hat{S} \to \hat{D}$ as limit map.

Proof. Clear. □

COROLLARY 4.3. Let $\{S_j \mid j \in I\}$ be a family in $\underline{S}$ and $\{T_j \mid j \in J\}$ the dual family in $\underline{Z}$ with $T_j = \hat{S}_j$. Then

(a) $S = \Pi S_j$ with projections $pr_j : S \to S_j$ iff $\hat{S} = \coprod T_j$ with $copr_j = pr_j\hat{} : \hat{S}_j \to \hat{S}$ as coprojections.

(b) $S = \coprod S_j$ with coprojections $\mathrm{copr}_j : S_j \to S$ iff $\hat{S} = \Pi \hat{S}_j$ with $\mathrm{pr}_j = \mathrm{copr}_j^{\wedge} : \hat{S} \to \hat{S}_j$ as projections.

Proof. Immediate from 4.2. □

REMARK. In terms of elements, this means that an element $(f_j)_{j \in J}$ of $\Pi \hat{S}_j$ may be identified with a character of $\coprod S_j$ such that for an element $(s_j)_{j \in j} \in \coprod S_j \subseteq \Pi S_j$ (where almost all s_j are 1) we have

$$(f_j)(s_j) = \inf\{f_j(s_j) \mid j \in J\}.$$

There are some general facts involving projectives and injectives which are of a general functorial nature. We discuss these in the following and defer a finer study of injectives and projectives to a later point (III-3).

It is useful to remark the nature of coproducts in $\underline{Z}$:

PROPOSITION 4.4. a) Let $\{S_j \mid j \in J\}$ be a family of semilattices in $\underline{Z}$. Then there is an injection (in $\underline{S}$)

$$\psi : \coprod_{\underline{S}} (S_j)_d \to \coprod_{\underline{Z}} S_j$$

(where $(S_j)_d$ is the underlying semilattice), such that the image is dense. In particular, a bijective copy of $\coprod_{\underline{S}} \hat{S}_j$ is a dense subset of $\coprod_{\underline{Z}} S_j$.

b) If all S_j in a) are finite, then

$$\coprod_{\underline{Z}} S_j \cong \coprod_{\underline{S}} S_j \quad \text{(See 3.7).}$$

Proof. a) We consider the following diagram

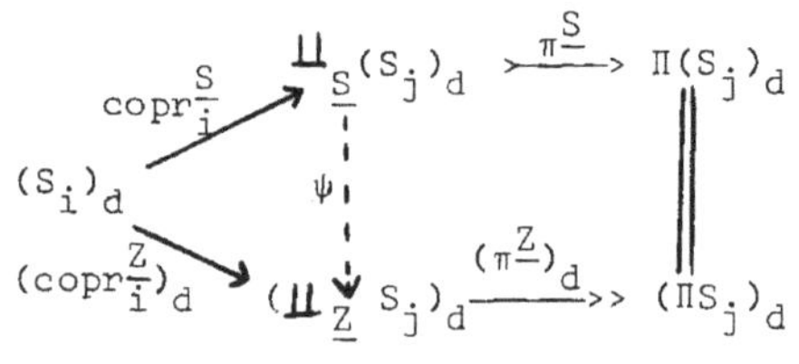

where π denotes the natural map from a coproduct into the product in a pointed category. Note that, since the image of $\pi^{\underline{S}}$ is dense in $\coprod S_j$, the image of $\pi^{\underline{Z}}$ is dense; hence $\pi^{\underline{Z}}$ is surjective. The $\underline{S}$-morphism ψ is given by the universal property of $\amalg_{\underline{S}}$. Since $(\pi^{\underline{Z}})_d\psi = \pi^{\underline{S}}$ is injective, then π is injective. We let $T = (\text{im}\ \psi)^-$. Then all coprojections $\text{copr}_i^{\underline{Z}}$ map into T, hence, by the universal property of $\amalg_{\underline{Z}}$, there is a unique $\underline{Z}$-morphism $\phi : \amalg_{\underline{Z}} S_j \to T$ with $\phi\ \text{copr}_j = \text{copr}_j'$ (the corestriction of copr_j to T). If $\phi' : T \to \amalg_{\underline{Z}} S_j$ is the inclusion, then $\phi'\text{copr}_j' = \text{copr}_j$, hence $\phi'\phi\ \text{copr}_j = \text{copr}_j$, whence $\phi'\phi = 1$ by the uniqueness of the fill-in map. Since ϕ' is an inclusion map we have $T = \amalg_{\underline{Z}} S_j$. We know that $K(S_j)$ is dense in S_j by II-1.9 and that there is a bijection between S_j and $K(S_j)$ (II-3.7). By the preceding, $\psi(\amalg_{\underline{S}} K(S_j))$ is then dense in $\amalg_{\underline{Z}} S_j$.

b) Since ^ is a left adjoint, it preserves coproducts. Thus $(\amalg_{\underline{S}}(S_j)_d)^\wedge \cong \amalg_{\underline{Z}}((S_j)_d)^\wedge$. If all S_j are finite, then $S_j = ((\overline{S}_j)_d)^\wedge$. □

DEFINITION 4.5. An object P in a category $\underline{A}$ is called projective for a class $\underline{E}$ of epics in $\underline{A}$ (shortly $\underline{E}$-projective) iff for each $e \in E$, $e : A \to B$ and each $f \in \underline{A}$, $f : P \to B$ there is an $f' : P \to A$ with $ef' = f$. We say that P is projective, if $\underline{E}$ is the class of all epics. An injective object is a projective in $\underline{A}^{op}$, the opposite category.

DEFINITION 4.6. Let $\underline{A}$ and $\underline{B}$ be categories; we say that $\underline{A}$ is $\underline{B}$-based iff there is a faithful functor $|\ | : \underline{A} \to \underline{B}$ which we will call the grounding functor. A category $\underline{A}$ is said to have a $\underline{B}$-free functor, if it is $\underline{B}$-based and the grounding functor has a left adjoint $F : \underline{B} \to \underline{A}$. All of this applies particularly to $\underline{B} = \text{Set}$, in which case we call F a free functor. An object then is free if it is of the form $F(X)$. A morphism in a $\underline{\text{Set}}$-based category is called injective or surjective iff $|f|$ is an injective, respectively surjective function.

Every surjective, resp. injective morphism is epic, resp. monic. We have noted that in the categories $\underline{S}$ and $\underline{Z}$ a morphism is epic (monic) iff it is surjective (injective)(see 1.6 and 2.8).

PROPOSITION 4.7. In a set-based category $\underline{A}$ every free object is Sur-projective, where Sur is the class of surjective maps.

Proof. Let $e : A \to B$ be surjective and $f : F(X) \to B$ be a morphism. Denote with $\phi_X : X \to |F(X)|$ the front adjunction. Since $|e| : |A| \to |B|$ is a surjective function of sets, then by the axiom of choice, there is a function $e' : |B| \to |A|$ with $|e|e' = 1_{|B|}$. Let $\psi : X \to |A|$ be defined by $\psi = e'|f|\phi_X$. Then $|e|\psi = |f|\phi_X$. Since F is a left adjoint there is a unique $f' : F(X) \to A$ such that $|f'|\phi_X = \psi$. Then $|ef'|\phi_X = |e|\psi = |f|\phi_X$. By the uniqueness in the universal property of the left adjoint, this implies $ef' = f$. □

LEMMA 4.8. Every retract of an E-projective is an E-projective.

Proof. Exercise: If P is E-projective and $\phi : P \to Q$ has a right inverse $\psi : Q \to P$, show that Q has the required property. □

PROPOSITION 4.9. In a set based category with a free functor, every Sur-projective is a retract of a free object.

Proof. Let P be a projective in $\underline{A}$ and let $\psi_A : F|A| \to A$ denote the back adjunction. Then, by one of the characteristic properties of adjunctions, the diagram

$$\begin{array}{ccc} & |F|A|| & \\ {}^{\phi_{|A|}}\nearrow & & \searrow^{|\psi_A|} \\ |A| & \xrightarrow{\quad 1_{|A|} \quad} & |A| \end{array}$$

commutes. In particular, ψ_P is surjective; hence since P is Sur-projective, there is a $\psi' : P \to F|P|$ such that $\psi_P\psi' = 1_P$. Thus P is a retract of $F|P|$. □

COROLLARY 4.10. In a set based category with a free functor, the following two statements are equivalent for an object P:

(1) P is Sur-projective (2) P is a retract of a free object.

Proof. 4.7 - 4.9. □

COROLLARY 4.11. In both categories $\underline{S}$ and $\underline{Z}$, an object is projective iff it is a retract (i.e. a direct factor) of a free object. □

Proof. 4.10, 1.5 and 2.7. □

PROPOSITION 4.12. Let $S \in \text{ob}\,\underline{S}$ and let $T \in \text{ob}\,\underline{Z}$ be its dual. Then

(a) S is projective iff T is injective,

(b) S is injective iff T is projective.

Proof. Clear from duality. □

COROLLARY 4.13. Let $S \in \underline{S}$. Then

(a) S is projective iff S is a retract (direct factor) of F(X) (the semilattice of finite subsets of X under $\cup$) for some set X.

(b) S is injective iff S is a retract (direct factor) of 2^X (the semilattice of all subsets of X under $\cap$) for some set X.

Proof. (a) is immediate from 4.11 and the structure of free semilattices (see 1.5).

(b) S is injective iff T is projective in $\underline{Z}$ iff T is a retract (direct factor) of some free object ${}^X2$ in $\underline{Z}$ (by 4.10) iff S is a coretract (direct factor) of $({}^X2)\hat{} \cong 2^X$ (by duality and 4.3).

COROLLARY 4.14. Let $T \in \underline{Z}$. Then

(a) T is projective (in $\underline{Z}$) iff $\hat{T}$ is injective in $\underline{S}$ iff T is a retract (topological direct factor) of some copower ${}^X2$ in $\underline{Z}$ for some set X.

(b) T is injective (in $\underline{Z}$) iff T is a retract (topological direct factor) of some 2^X for some set X.

Proof. Clear from 4.12 by duality. □

REMARK. In $\underline{Z}$, by 4.4, the copower ${}^X2$ is the Bohr compactification of the copower of X copies of 2 in $\underline{S}$, which is F(X), i.e. ${}^X2$ (in $\underline{Z}$) is isomorphic to $\alpha(F(X))$ (in the terminology of 3.11, where F(X) has the discrete topology).

Clearly, the characterization of the projectives in $\underline{Z}$ so far is the least satisfactory, since the nature of the free object X_2 in $\underline{Z}$ is still somewhat obscure. It is therefore useful to add yet another aspect of the free object $F_{\underline{Z}}(X)$ in $\underline{Z}$. By 2.6 we know that the $\underline{ZComp}$-free object over a compact zero dimensional space Y is $\Gamma(Y)$, the space of all compact subsets of Y with $\cup$ as semilattice operation. The grounding functor $\underline{ZComp} \to \underline{Set}$ has as its left adjoint the Čech compactification functor $\beta : \underline{Set} \to \underline{ZComp}$. Since left adjoints compose, we have X_2 (in $\underline{Z}$) $= F_{\underline{Z}}(X) \cong \Gamma(\beta X)$. Thus

COROLLARY 4.15. T is projective in $\underline{Z}$ iff it is a retract (topological direct factor) of $\Gamma(\beta X)$ for some set X. □

DEFINITION 4.16. A topological space is called extremally disconnected iff the closure of every open set is open.

A compact space Y is extremally disconnected iff the Boolean lattice $\Gamma(Y)$ of open closed sets is complete. For any discrete space X the Čech compactification of X is extremally disconnected [G-2].

PROPOSITION 4.17. A $\underline{Z}$-object T is projective in $\underline{Z}$ iff there is an extremally disconnected compact space E such that T is a retract (topological direct factor) of $\Gamma(E)$.

Proof. By 4.15 and the preceding remarks the necessity follows with $E = \beta X$. Conversely, assume that T is a retract of $\Gamma(E)$ for some extremally disconnected compact space E. Let $|E|$ be the underlying set and let $\mu : \beta|E| \to E$ be the back adjunction for β. By Gleason's theorem, the extremally disconnected spaces are the projectives in the category of compact spaces (see [G-3]). Hence there is a continuous function $\gamma : E \to \beta|E|$ such that $\mu\gamma = 1_E$. Then $\Gamma(\mu)\Gamma(\gamma) = 1_{\Gamma(E)}$, showing that $\Gamma(E)$ hence also T is a retract in $\underline{Z}$ of the free object $F_{\underline{Z}}(|E|) = \Gamma(\beta|E|)$. Hence T is projective by 4.15. □

Historical Notes for Chapters 0 and I.

The results of this chapter in themselves are not really new. As mentioned in the introduction, the forerunner for a character and duality theory for semilattices was Austin [A-2]. The theory was expanded by Schneperman [S-4] and by Baker and Rothman [B-1] to cover locally compace situations. Recent contributions are due to Bowman [B-10]. The precise statement of our duality in full generality together with the relation to the category in exercise 3.12 was discussed by Hofmann [H-4]. The proof we presented is new, although the style of the proof is parallel to a recent proof of the duality theorem for locally compact abelian groups by Roeder [R- 3]. Both proofs rely on density and functorial continuity arguments which have become a whole special interest area of category theory. The set-up is modelled after that used previously by Hofmann [H-3] for certain applications in compact group theory. Our category $\underline{S}$ of semilattices (with identity) is a prime example for a semiadditive balanced complete and cocomplete category, which is not additive. An abstract characterization of this category does not seem available. A very general theory covering the category theoretical background far beyond what is needed here has recently been developed by Isbell [I-2].

CHAPTER II. The character theory of compact and discrete semilattices

While in Chapter I we introduced the duality of the category $\underline{Z}$ of compact zero dimensional semilattices as a special case of a very general theory which functions under a great variety of category theoretical circumstances, we will emphasize in this Chapter the specific features of this duality. These particular characteristics pertain to the mathematics of discrete and topological semilattices and (with a few exceptions) are absent in parallel theories such as the Pontryagin duality between compact abelian and abelian groups. It is these special features which play a crucial role in the applications of the duality theory.

We will practice the following notational convention: If S is a semilattice and $\{s_1,\ldots,s_n\} \subseteq S$ we write $\wedge\{s_1,\ldots,s_n\}$ for $s_1\ldots s_n$. For arbitrary subsets $X \subseteq S$ we write $\inf X$ and $\sup X$ for the greatest lower and the least upper bound.

Section 1. The category of zero dimensional compact semilattices

In this section we treat compact zero dimensional semilattices as particular types of compact monoids and utilize their general theory

We first prove the monotone convergence theorem in compact topological semilattices.

Recall that a net in a set X is a function $J \to X$ where J is an upwards directed set. If S is a semilattice, a net $n \mapsto x_n : J \to S$ is monotone if $m \le n$ in J implies $x_m \le x_n$ (resp. $x_n \le x_m$) for all $m,n \in J$. It is useful to prove some of our results for a class of semilattices properly containing the class of topological semilattices. We recall that a monoid S is called semitopological if it carries a topology relative to which all translations $s \mapsto sa$, $s \mapsto as : S \to S$, $a \in S$ are continuous. Of course every topological semilattice is semitopological.

PROPOSITION 1.1. Let S be a compact semitopological semilattice. Then any monotone net in S converges. In fact, if $(x_n)_{n\in J}$ is a decreasing (resp. increasing) net, then $\lim_J x_n = \inf\{x_n \mid n \in J\}$ (resp., $\lim_J x_n = \sup\{x_n \mid n \in J\}$.

Proof. Assume that $(x_n)_{n\in J}$ is decreasing and let a and b be cluster points of this net. For fixed $j \in J$ and all $n \in J$ with $j \leq n$ we have $x_n \leq x_j$, i.e. $x_n x_j = x_n$. The continuity of the translations implies $ax_j = a$. Since j is arbitrary, we conclude $ab = a$ by the continuity of translations. By symmetry we have $ab = b$. Hence $x = \lim x_n$ exists. From $xx_j = x$ for all j we conclude $x \leq x_j$ for all $j \in J$, i.e. x is a lower bound for $\{x_n \mid n \in J\}$. Assume that y is any lower bound. Then $yx_n = y$ for all n, implying $yx = y$ by continuity of translations. Hence $y \leq x$, i.e. $x = \inf\{x_n \mid n \in N\}$. Now assume that $(x_n)_{n\in J}$ is increasing. Then, for fixed $j \in J$ and all $n \in J$ with $j \leq n$ we have $x_j \leq x_n$, i.e. $x_j = x_j x_n$. If a is a cluster point, by the continuity of translations we conclude $x_j = x_j a$, and then $b = ba$ for any other cluster point b. By symmetry again $a = ab = b$, and $x = \lim x_n$ exists. As before, x is an upper bound of $\{x_n \mid n \in J\}$, and if y is any upper bound, then $x_n = x_n y$ yields $x = xy$, hence $x = \sup\{x_n \mid n \in J\}$. □

LEMMA 1.2. If S is a semilattice such that $\inf X$ exists for each $X \subseteq S$, then $\sup X$ exists for each $X \subseteq S$ and equals $\inf\{b \in S \mid b$ is an upper bound of $X\}$. In particular, S is a complete lattice with zero. (Recall that "semilattice" implies the existence of an identity according to our Definition I-1.1).

Proof. Standard. □

COROLLARY 1.3. Let S be a compact semitopological semilattice. Let X be an arbitrary subset of S and J the $\subseteq$-directed set of finite subsets F of X. Then $\inf X$ and $\sup X$ exist. Moreover $\inf X = \lim_J \inf F$,

$\sup X = \lim_J \sup F = \inf\{b \in S \mid b$ is an upper bound of $X\}$,

Proof. The net $(x_F)_{F \in J}$, $x_F = \wedge F$ is decreasing. Hence by 1.11, $\lim_J x_F = \inf\{x_F \mid F \in J\}$ exists. The left hand side clearly agrees with $\inf X$, so the first part of the assertion follows. The remainder follows from 1.1 and 1.2. □

COROLLARY 1.4. Any compact semitopological semilattice S is a complete lattice; in particular, S has a zero. □

The following is a characterization of Z-objects among all compact zero dimensional spaces which have a semilattice structure:

THEOREM 1.5. Let S be a semilattice and a compact zero dimensional space. Then the following statements are equivalent:

(1) S is a semitopological semilattice (i.e. all translations $s \mapsto sa: \longrightarrow S$ are continuous).

(2) S is a Z-object, i.e. multiplication $(s,t) \mapsto st : S \times S \to S$ is continuous.

(3) S is profinite (i.e. is the projective limit of finite semilattices).

(4) The continuous characters $S \to 2$ separate the points.

These conditions imply

(5) Each point has a neighborhood basis of open closed proto-subsemilattices.

Proof. (1) => (2) is a recent result of Lawson's [L-2]. (2) => (3) holds in fact for every compact zero dimensional topological monoid (Numakura [N-2]). (3) => (4) Clear since the characters separate on finite semilattices. (4) => (5) and (2). Let $\hat{S}$ denote the set of continuous characters. By (4) there is an embedding and semilattice injection $S \to 2^{\hat{S}}$. Clearly $2^{\hat{S}}$ is topological and has arbitrarily small open closed sets which are closed under multiplication, hence so does S, and S is topological. Finally, (2) => (1) is trivial. □

Two concepts are fundamental in the whole theory which

we propose, namely those of a local minimum and local maximum. It is feasible to introduce these concepts in general quasi-ordered spaces.

DEFINITION 1.6. Let S be a set with quasi-order $\leq$. We define

(i) $\uparrow s = \{x \in S \mid s \leq x\}$

(ii) $\downarrow s = \{x \in S \mid x \leq s\}$.

These sets are called the upper and the lower set (and they are sometimes denoted by $M(s)$ and $L(s)$ in the literature). In accordance with the use in semigroup theory we write $S \setminus \uparrow s = I(s)$.

DEFINITION 1.7. Let S be a set with a quasi-order and a topology. An element $s \in S$ is called

(i) a local minimum [local maximum] iff there is an open neighborhood of s with $\downarrow s \cap U \subseteq \uparrow s$ [$\uparrow s \cap U \subseteq \downarrow s$]

(ii) a strong local minimum [maximum] iff $\uparrow s$ [resp., $\downarrow s$] is open,

(iii) a semi-minimum iff s is a minimum of $S \setminus \downarrow t$ for some strong local maximum t.

The property of being a semimaximum is defined dually.

LEMMA 1.8. Let S be a commutative monoid and $\leq$ the quasi-order defined by $s \leq t$ iff $s \in St$. Write $H(s)$ for $\uparrow s \cap \downarrow s$. Then

(a) $\downarrow s = Ss$ for all $s \in S$,

(b) $\uparrow s = \{x \in S \mid xs \in H(s)\}$ for all idempotents $s \in S$,

(c) $I(s)$ is the unique largest ideal not containing s.

Note that $\uparrow s$ in (b) is a submonoid.

Proof. (a) is trivial. (b) It is readily seen that $H(s)$ is a group with identity s if s is idempotent. If $xs \in H(s)$, then $xs = h$ with a group element $h \in H(s)$, hence $xsh^{-1} = hh^{-1} = s$, i.e. $s \leq x$. Conversely if $s \leq x$, then $s = xy = x(ys)$ since s is idempotent, but $s \leq y$ implies $s = s^2 \leq ys \leq s$, i.e. $ys \in H(s)$. (c) is straightforward, since $I(s) = \cup\{Sx \mid s \nleq x\}$. □

COROLLARY 1.9. Let S be a commutative semitopological monoid with the quasi-order of 1.8. For an idempotent $e \in S$ the following statements are equivalent:

(1) e is a local minimum.

(2) $H(e)$ is open in Se.

(3) $\uparrow e$ is open, i.e. e is a strong local minimum.

(4) $I(e)$ is closed.

Proof. (1) => (2): Let U be open such that $e \in U \subseteq Se \subseteq \uparrow e$. Then $U \cap Se \subseteq \uparrow e \cap Se = H(e)$. Thus $U \cap Se$ is an open subset of Se which is contained in the group of units $H(e)$ of Se. Since the translations of Se by elements of $H(e)$ have inverses, they are homeomorphisms of Se. Thus $H(e) = \cup\{h(U \cap Se) \mid h \in H(e)\}$, whence $H(e)$ is open in Se. (2) => (3): Let $r : S \to Se$ be defined by $r(s) = se$. Then $\uparrow e = r^{-1}(H(e))$ by 1.8. Hence $\uparrow e$ is open. (3) <=> (4): trivial. (3) => (1) Take $U = \uparrow e$. □

We will use this result in the following form:

PROPOSITION 1.10. Let S be a semitopological semilattice. Then for any $s \in S$ the following statements are equivalent:

(1) s is a local minimum.

(2) s is a strong local minimum.

(3) s is isolated in Ss.

(4) $\uparrow s$ is open closed.

(5) $I(s)$ is open closed.

Moreover, if s and t are local minima, then so is $\sup\{s,t\}$.

Proof. We need to show that $\uparrow s$ is always closed, but $\uparrow s$ is the inverse image of the closed set $H(s) = \{s\} \subseteq Ss$ under the map $x \to xs$. The last assertion follows from $\uparrow \sup\{s,t\} = \uparrow s \cap \uparrow t$. □

DEFINITION 1.11. In a semi-topological semilattice S, the sup-semilattice of all local minima will be denoted by $K(S)$.

The following is a rather crucial collection of results on local minima:

<u>THEOREM 1.12</u>. Let S be a compact zero dimensional

topological semilattice. Then we have:

(1) $K(S) \cap Ss = K(Ss)$ for each $s \in S$.

(2) $\sup(K(S) \cap Ss) = s$ for each $s \in S$.

(3) $K(S)$ is dense in S.

Proof. (1) If $k \in K(S) \cap Ss$, then $\uparrow k \cap Ss$ is open, so $k \in K(Ss)$. Conversely, if $k \in K(Ss)$, then $\uparrow_{Ss} k$ is open in Ss. Let $f : S \to S$ be the continuous map given by $r(x) = xs$; then $\uparrow_S k = r^{-1}(\uparrow_{Ss} k)$. Hence $\uparrow k$ is open and $k \in K(S)$.

(2) By (1) it suffices to show that $\sup K(S) = 1$. By 1.5 there are arbitrarily small open closed subsemilattices U around 1. If $k = \min U$, then $Sk \cap U = \{k\}$, hence $k \in K(S)$ by 1.10. If $s \in S$, $s \neq 1$, then Ss is closed, and if U is chosen so that $Ss \cap U = \emptyset$, then $k \not\leq s$. Hence $\sup K(S) = 1$.

(3) follows from (2) and 1.3.

EXAMPLE 1.13. Let $\rho : C \to I$ be the Cantor function from the standard Cantor subsemilattice C of the unit interval semilattice I which maps $\{1/3, 2/3\}$ to $1/2$, $\{1/9, 2/9\}$ to $1/4$, $\{7/9, 8/9\}$ to $3/4$, etc. Let S be the quotient of $C \times C$ modulo the congruence relation whose cosets are $\{0\} \times \rho^{-1}(t)$, $t \in I$ and $\{(x,y)\}$ for $x > 0$. Then S is one dimensional but the set of local minima is still dense in S. □

HISTORICAL REMARK. 1.12 and 1.13 were (partly) introduced in [H-9]; the example 1.13 is the mapping cylinder Cyl(C,ρ) in terms of [H-8]. □

There is always an abundance of semimaxima, too.

PROPOSITION 1.14. Let $S \in$ ob $\underline{Z}$. Then

$$s = \inf\{t \mid t \text{ is a semimaximum and } s \leq t\}$$

for all $s \in S$.

Proof. By 1.4 we can define $s' = \inf\{t \mid s \leq t,\ t$ semimaximum$\}$. Clearly $s \leq s'$. Assume $s < r \in S$. By 1.12 we find a $k \in K(S)$ with $k \leq r$ and $s \in I(k)$. Let t be a maximal element of $I(k)$ dominating s; such exists since $I(k)$ is closed by 1.10. Then t is a semimaximum, and thus $s' \leq t$ by definition of s'. Hence $s' \leq t \in I(k)$ and $k \leq r$ imply $r \neq s'$. Therefore $s = s'$. □

COROLLARY 1.15. The subsemilattice generated in a $\underline{Z}$-object by the semimaxima is dense.

Proof. 1.3 and 1.14. □

HISTORICAL COMMENT. Proposition 1.16 in the form of an algebraic analogue is due to Dilworth and Crawley [D-3].

By contrast with the local minima in a $\underline{Z}$-object, the local maxima are not generally strong local maxima.

EXAMPLE 1.16. Let $T = \{0\} \cup \{1/n \mid n = 1,2,\ldots\}$ under min, and $T' = \{0,a,b\}$ with $ab = 0$. Let $S = (T \times T')/(T \times \{0\} \cup \{0\} \times T') \cup \{1\}$ (with 1 as isolated identity. Then S has no strong local maxima other than 1.

REMARK 1.17. In a compact topological commutative monoid S with the quasi-order of 1.8 an idempotent is a local maximum iff $H(e)$ is open in $\uparrow s$.

Proof. Exercise. □

LEMMA 1.18. In a topological semilattice, the set of strong local maxima is a subsemilattice.

Proof. Since $S1 = S$ is open, 1 is a strong local maximum. If Ss and St are open, so is $Sst = Ss \cap St$. □

DEFINITION 1.19. Let $S \in \text{ob}\ \underline{Z}$. The subsemilattice of strong local maxima will be denoted with $K_{co}(S)$. From example 1.16 we know that $K_{co}(S) = \{1\}$ is quite possible. The significance of the strong local maxima, to which we will return later, is that they will allow us to characterize those $\underline{Z}$-objects which are, in effect, topological lattices (i.e. have continuous sup-operation).

Section 2. Characters and filters on discrete semilattices.

In this section we relate the character theory of discrete semilattices to the more conventional idea of filters. The character semilattice of a discrete semilattice may so be interpreted as a semilattice of filters with a suitable topology.

DEFINITION 2.1. A subset F of a semilattice is called a filter iff it is a subsemilattice and the conditions $f \in F$ and $f \leq s$ imply $s \in F$. A filter F with a smallest element s is called principal, and s is called the generator of F. The set of all filters on S will be denoted $\mathcal{F}(S)$. □

We observe that the principal filter generated by s is precisely $\uparrow s$. We recall that an ideal I in a commutative semigroup is a subset satisfying $SI \subseteq I$.

DEFINITION 2.2. An ideal I in a commutative semigroup is a prime ideal iff $S \setminus I$ is a subsemigroup. □

The connection between character theory and these concepts is given in the following

PROPOSITION 2.3. Let S be a semilattice (i.e. an $\underline{S}$-object) and let $f : S \to 2$ be a function. Then the following statements are equivalent:

(1) f is a character (i.e. $f \in \hat{S}$).

(2) $f^{-1}(1)$ is a filter (i.e. $f^{-1}(1) \in \mathcal{F}(S)$).

(3) $f^{-1}(0)$ is a prime ideal.

Proof. (1) => (2) is immediate from the definitions.

Suppose that (2) is satisfied. Then $I = f^{-1}(0)$ is the complement of a filter F. Suppose that $i \in I$ and $s \in S$. If $si \notin I$, then $si \in F$. Then $si \leq i$ implies $i \in F$ by 2.1. But $i \in I = S \setminus F$. This contradiction shows that $si \in I$. Hence I is an ideal. Since $S \setminus I = F$ is a subsemigroup, I is a prime ideal. Since $F \neq \emptyset$, then I is proper. Finally assume (3). Suppose that $s,t \in S$. If $f(s) = f(t) = 1$, then $f(st) = 1$, since $f^{-1}(1)$ is a subsemigroup by (3). If $f(t) = 0$, then $f(st) = 0$ since $f^{-1}(0)$ is an ideal. Hence f is a morphism. □

HISTORICAL REMARK. The relations between filters, prime ideals and characters were widely exploited in [H-5].

The connection between $\mathcal{F}(S)$ and the character semilattice is finally elucidated in the following:

PROPOSITION 2.4. i) If S is a semilattice, then the set

$\mathfrak{F}(S)$ of filters on S is a semilattice relative to $\cap$. The sets $W(s) = \{F \in \mathfrak{F}(S) \mid s \in F\}$ together with the sets $\mathfrak{F}(S)\backslash W(s)$, $s \in S$ form a subbasis for a compact zero dimensional topology on $\mathfrak{F}(S)$ making $\mathfrak{F}(S)$ into a $\underline{Z}$-object.

ii) The function $\phi_S : \hat{S} \to \mathfrak{F}(S)$ which is given by $\phi_S(f) = f^{-1}(1)$ is an isomorphism of $\underline{Z}$-semilattices. Its inverse is given by $\phi_S{}^{-1}(F)(s) = 1$ if and only if $s \in F$ (and $= 0$ otherwise).

iii) If $f : S \to T$ is an $\underline{S}$-morphism, then $f^{-1}(G) \in \mathfrak{F}(S)$ for each $G \in \mathfrak{F}(T)$. The function $\mathfrak{F}(f) : \mathfrak{F}(T) \to \mathfrak{F}(S)$ given by $\mathfrak{F}(f)(G) = f^{-1}(G)$ is an $\underline{S}$-morphism, and the diagram

$$\begin{array}{ccc} \hat{T} & \xrightarrow{\hat{f}} & \hat{S} \\ {\scriptstyle \phi_S}\downarrow & & \downarrow{\scriptstyle \phi_S} \\ \mathfrak{F}(T) & \xrightarrow[\mathfrak{F}(f)]{} & \mathfrak{F}(S) \end{array}$$

commutes. In particular, $\mathfrak{F} : \underline{S} \to \underline{Z}^{op}$ is a functor and $\phi : \hat{} \to \mathfrak{F}$ is a natural isomorphism of functors.

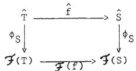

Proof. Clearly the intersection of two filters is again a filter and $S \in \mathfrak{F}(S)$. Thus $\mathfrak{F}(S)$ is a $\cap$-subsemilattice of the $\cap$-semilattice of all subsets of S. If $f : S \to T$ is a morphism and G a filter on T, then $f^{-1}(G)$ is a subsemilattice and satisfies the convexity condition required for a filter (2.1). In particular, if $T = 2$ and $G = \{1\}$, then $f^{-1}(1) \in \mathfrak{F}(S)$ follows. Hence ϕ_S is well-defined. If $f,g \in S$, then $\phi_S(fg) = (fg)^{-1}(1) = f^{-1}(1) \cap g^{-1}(1) = \phi_S(f)\phi_S(g)$, so ϕ_S is a morphism. Since the function $\mathfrak{F}(S) \to \hat{S}$ which associates with a filter F the character taking the value 1 precisely on F is obviously an inverse for ϕ, then ϕ_S is an isomorphism of semilattices. It maps the set $W_1(s) = \{g \in \hat{S} \mid g(s) = 1\}$ onto $W(s)$, and the set $W_0(s) = \{g \in \hat{S} \mid g(s) = 0\}$ onto $S\backslash W(s)$; since the topology on S is that of pointwise convergence, the sets $W_0(s)$, $W_1(s)$ generate the topology of $\hat{S}$. Hence $W(s)$ and $S\backslash W(s)$ generate on $\mathfrak{F}(S)$ a topology relative to which ϕ_S is a homeomorphism. This finishes i) and ii) completely.

It remains to show that the diagram in iii) is commutative. Now $f(g) = g \circ f$, hence $(\phi_S f)(g) = (g \circ f)^{-1}(1)$. On the other hand, $\mathfrak{F}(f)[\phi_T(g)] = \mathfrak{F}(f)[g^{-1}(1)] = f^{-1}[g^{-1}(1)] = (g \circ f)^{-1}(1)$, proving the assertion. □

In the future we will denote with $\mathfrak{F}(S)$ the $\underline{Z}$-object defined by 2.4 and with $\mathfrak{F}$ the contravariant functor $\underline{S} \to \underline{Z}^{op}$ given by 2.4, which is naturally isomorphic to ^. By way of an application, we observe the following

PROPOSITION 2.5. Let $S \in$ ob S. The function $\beta_S : S \to \mathfrak{F}(\mathfrak{F}(S)_d)$ which associates with each $s \in S$ the set $W(s)$ of all $F \in \mathfrak{F}(S)$ with $s \in F$ is equivalent to the Bohr compactification of S. Specifically, the diagram

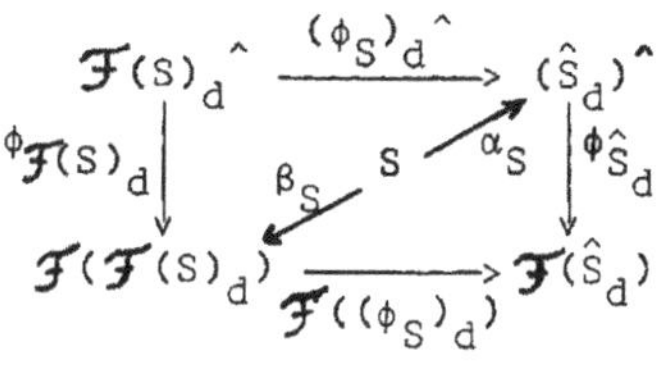

is commutative.

Proof. Exercise. □

Section 3. The algebraic characterization of the category $\underline{Z}$.

This section is perhaps the most important portion of this chapter as far as the applications are concerned. It brings the ideas of the preceding two sections together and yields further structural information about compact zero dimensional semilattices and obtains their characterization in purely lattice theoretical terms.

The following is a standard concept in algebraic lattice theory:

DEFINITION 3.1. An element k in a semilattice is called compact iff $k \leq \sup X$ for any subset $X \subseteq S$ implies the existence of a finite subset $F \subseteq X$ with $k \leq \sup F$.

LEMMA 3.2. If c and k are compact elements in a semilattice and if $\sup\{c,k\}$ exists, then $\sup\{c,k\}$ is compact.

Proof. Exercise. □

We now present the crucial theorem of the section which characterizes the compact elements in any Z-object.

THEOREM 3.3. Let S be a compact zero dimensional topological semilattice. Then for any element $k \in S$ the following statements are equivalent:

(1) k is compact.

(2) $k \in K(S)$ (i.e. k is a local minimum, see 1.7, 1.10).

(3) $\uparrow k$ is an open closed principal filter.

(4) $I(k)$ is an open closed prime ideal.

(5) k is isolated in Sk.

(6) The function $f_k : S \to 2$ defined by $f_k(s) = 1$ iff $k \leq s$ (and $= 0$ otherwise) is an element of $\hat{S}$.

(7) $\{f \in \hat{S} \mid f(k) = 1\}$ is a principal filter of $\hat{S}$ (which is generated by f_k of (5)).

(8) For each chain $Y \subseteq Sk \setminus \{k\}$, $\sup Y = \lim Y < k$.

Proof. By 1.10 and 2.3 we know that (2) through (6) are equivalent. Now assume (1) and set $X = K(S) \cap Sk$. Then $k = \sup X$ by 1.12. Since k is compact, there is a finite subset $F \subseteq K(S) \cap Sk$ with $k \leq \sup F \leq \sup X = k$ by 3.1. By 1.10 we have $\sup F \in K(S)$, hence $k \in K(S)$, so (2) holds. Suppose (3) and assume that $k \leq \sup X$ for some $X \subseteq S$. Let J be the $\subseteq$-directed set of finite subsets of X. Then $\sup X = \lim_J \sup F$ by 1.3. Since $\uparrow k$ is a neighborhood of k by (3) there is an $F \in J$ with $\sup F \in \uparrow k$, i.e. $k \leq \sup F$. This shows that k is compact. Hence (3) $\Rightarrow$ (1). (6) $\Rightarrow$ (7) is clear and we now prove (7) $\Rightarrow$ (5). For this purpose we apply duality and 2.4 and assume that $S = \mathcal{F}(T)$ (so that $T \cong \hat{S}$); then k is a filter on T. We may identify a character f of S with an element $t = t_f \in T$ in such a fashion that $f(F) = 1$ iff $t \in F$ for $F \in S = \mathcal{F}(T)$. Under this identification, $\{f \in S \mid f(k) = 1\}$ becomes identified with k. Thus we assume that k is a principal filter on T and we show that k is isolated in $Sk = \{F \in \mathcal{F}(T) \mid F \subseteq k\}$. Since k is principal, then $k = \uparrow t$ for some $t \in T$. By 2.4, the set $W(t) = \{F \in \mathcal{F}(T) \mid t \in F\}$ is open in S, hence a neighborhood of $k = \uparrow t$. However, if $F \in k \cap W(t)$, then $F \subseteq k = \uparrow t$ and $t \in F$, hence $F = \uparrow t$. Thus $Sk \cap W(t) = \{k\}$ which had to be shown.

Clearly (1) implies (8). For the converse, suppose $X \subseteq Sk\backslash\{k\}$ and $\sup F < k$ for each $F \subseteq X$ finite. Then, $D = \{\sup F : F \subseteq X \text{ finite}\} \subseteq Sk\backslash\{k\}$ is directed by $d \leq d'$ if and only if $dd' = d$. Consider the following

Lemma(Iwamura). Let Y be a partially ordered set in which each directed subset has a sup. Let $Z \subseteq Y$ such that each non-empty chain $C \subseteq Z$ satisfies $\sup C \in Z$. Then $\sup D \in Z$ for each directed subset D of Z. (For a proof, see [M-1], S. 238, Hilfsatz).
If we take $S = Y$, $Sk\backslash\{k\} = Z$, and $D = D$, then the lemma shows $\sup D \in Sk\backslash\{k\}$, i.e. $\sup D < k$. Since $\sup D = \sup X$, we have (1). □

In view of 1.12 we can formulate the

COROLLARY 3.4. In a $\underline{Z}$-object S the set of compact elements is dense and each element of S is the l.u.b. of all compact elements which it dominates. □

The following is classical terminology which allows us to express this fact more smoothly.

DEFINITION 3.5. A lattice L is called an algebraic lattice if

(i) L is a complete lattice,

(ii) every element s is the l.u.b. of the set of all compact elements which it dominates (i.e. $s = \sup\{k \in S \mid k \text{ compact and } k \leq s\}$ for all s).

COROLLARY 3.6. If S is a compact zero dimensional semitopological semilattice, then the underlying semilattice is in fact an algebraic lattice.

Proof. 1.4 and 3.4. □

In Section 2 we found an alternative description of the dual semilattice of a discrete semilattice; we have now within our grasp an extremely useful alternative representation of the dual semilattice of a $\underline{Z}$-object:

THEOREM 3.7. Let $S \in \text{ob}\ \underline{Z}$. Then the function $\kappa_S : \hat{S} \to K(S)$ given by $\kappa_S(f) = \inf f^{-1}(1)$ is an isomorphism of semilattices, whose inverse is given by $k \mapsto f_k : K(S) \to S$ (with f_k as in 3.3 (5)).

Proof. If $f \in S$, then $f^{-1}(1)$ is an open closed filter. Since it is compact, it has a zero by 1.4, and by 3.3 this zero belongs to $K(S)$, since $f^{-1}(1)$ is open. But this zero is $\inf f^{-1}(1)$. Hence the function κ_S is well defined. It is straightforward that κ_S is a morphism since $\kappa_S(fg) = \inf[(fg)^{-1}(1)] = \inf[f^{-1}(1) \cap g^{-1}(1)] = (\inf f^{-1}(1)) \vee (\inf g^{-1}(1)) = \kappa_S(f) \vee \kappa_S(g)$. If $g \in \hat{S}$, then $f_{\kappa_S(g)}(s) = 1$ iff $\kappa_S(g) \leq s$ iff $\inf g^{-1}(1) \leq s$ iff $g(s) = 1$, whence $f_{\kappa_S(g)} = g$. Conversely, let $k \in K(S)$. Then $\kappa_S(f_k) = \inf f_k^{-1}(1) = k$. Therefore $k \mapsto f_k : K(S) \to \hat{S}$ is indeed the inverse function of κ_S.□

The following is a further consequence of 3.3.

PROPOSITION 3.8. Let S be a semilattice. Then a filter $F \in \mathfrak{F}(S)$ is a compact element in $\mathfrak{F}(S)$ (see Section 2) iff it is principal.

Proof. We may identify S with $\mathfrak{F}(S)\hat{}$ by associating with $s \in S$ the character χ_s of $\mathfrak{F}(S)$ which is defined by $\chi_s(G) = 1$ iff $s \in G$. Then F becomes identified with $\{\chi \in \mathfrak{F}(S)\hat{} \mid \chi(F) = 1\}$. Then (1) <=> (7) in 3.3 proves the proposition. □

We will now show that 3.6 has a complete inverse, i.e. that on each algebraic lattice we find a unique zero dimensional compact topology relative to which it is a topological semilattice such that the compact elements are the local minima.

PROPOSITION 3.9. Let S be a semilattice and K(S) the sup-semilattice of compact elements. Define

$\gamma = \gamma_S : S \to \mathfrak{F}(K(S))$ by $\gamma(s) = K(S) \cap Ss$. Then the following conclusions hold:

(i) γ is a morphism of semilattices.

(ii) If S is compactly generated (i.e. if $s = \sup K(S) \cap Ss$ for all s), then γ is injective.

(iii) If S is a complete lattice, then γ is surjective.

Proof. (i) The set $K(S) \cap Ss$ is a filter in $K(S)$, hence γ is well-defined. We have $\gamma(s)\gamma(t) = (K(S) \cap Ss) \cap (K(S) \cap St) = K(S) \cap Ss \cap St = K(S) \cap Sst = \gamma(st)$.

(ii) Suppose $\gamma(s) = \gamma(t)$. Then $K(S) \cap Ss = K(S) \cap St$. If S is compactly generated then $s = \sup(K(S) \cap Ss) = \sup(K(S) \cap St) = t$.

(iii) In order to show the surjectivity of γ, we take any filter F on $K(S)$ and show that it is of the form $K(S) \cap Ss$ for some s. Since S is a complete lattice, $s = \sup F$ exists in S. Then $F \subseteq K(S) \cap Ss$. Now take an arbitrary $k \in K(S) \cap Ss$. Then $k \leq s = \sup F$, so there is a finite subset $E \subseteq F$ with $k \leq \sup E$ since k is compact. Since F is a proto-submonoid of $K(S)$ and E is finite, we have $\sup E \in F$, and since F is a filter, this implies $k \in F$. Thus $K(S) \cap Ss \subseteq F$, showing the desired equality. □

<u>THEOREM 3.10</u>. Let S be an algebraic lattice. Then there is a compact zero dimensional topology on S making S into a topological semilattice such that the compact elements of S are precisely the local minima and this topology is the only compact semilattice topology with this property.

Proof. By Lemma 3.10 there is an isomorphism $\gamma : S \to \mathfrak{F}(K(S))$, where $K(S)$ is the sup semilattice of compact elements in S. Now $\mathfrak{F}(K(S))$ has a compact zero dimensional semilattice topology relative to which the local minima are precisely the principal filters (2.4, 3.3 and 3.9). But $\gamma(s) = K(S) \cap Ss$ is a principal filter in $K(S)$ iff $s = \sup \gamma(s) \in K(S)$. Hence S has a compact zero dimensional semilattice topology τ relative to which $K(S)$ is the set of local minima. Relative to any compact topology σ making S into a semitopological semilattice for which the compact elements are the local minima, the sets $\uparrow k$ have to be open closed for all $k \in K(S)$. But these sets and their complements generate τ by 2.4. Hence $\tau \subseteq \sigma$, and since σ is compact, the topologies have to agree. □

DEFINITION 3.11. The topology introduced in 3.9, 3.10 on an algebraic lattice is called the $\underline{Z}$-topology. □

REMARK. By a result of Lawson [L-1], on a given semilattice there is at most one compact Hausdorff topology making it into a topological semilattice. Hence the $\underline{Z}$-topology is in fact the only possible semilattice topology.

In the light of 3.6 and 3.10 we could say that compact zero dimensional (semi) topological semilattices and algebraic lattices are one and the same thing.

This is a very good point to recognize the connection of our presentation with some standard concepts and constructions in lattice theory. Let S be a semilattice and $T = \hat{S}$ its dual. By 3.7 we identified S with the sup semilattice $K(T)$, and by 3.9 we have identified T with the filter semilattice of the sup-semilattice $K(T)$. We should emphasize here that what we call a filter on a semilattice when applied to a sup-semilattice such as $K(T)$ is what also has been called in the literature a lower set, or lower end, or in fact most frequently, an ideal of the sup semilattice. Consequently, the underlying lattice of T appears as Nachbin's ideal completion of the sup semilattice $K(T)$ which, we reiterate, as a semilattice is isomorphic to S. For further references on the ideal completion see [G-4, M-4, N-1, S-3].

In the earlier part of the section we identified the local minima in a $\underline{Z}$-object algebraically as the compact elements and lead to an algebraic characterization of all $\underline{Z}$-objects. We now propose to carry out an analogous discussion for the strong local maxima (1.8, 1.15-1.18).

DEFINITION 3.12. An element m in a semilattice S is called cocompact iff $\inf X \leq m$ for any subset $X \subseteq S$

implies the existence of a finite subset $F \subseteq X$ with $\inf F \leq m$. Obviously 1 is always cocompact.

LEMMA 3.13. If m,n are cocompact elements in a semilattice, then mn is cocompact. (I.e., the set of cocompact elements is a subsemilattice.)

Proof. Exercise. □

PROPOSITION 3.14. Let S be a $\underline{Z}$-object. Then for any element $m \in S$ the following statements are equivalent:

(1) $m \in K_{co}(S)$, i.e. Sm is an open (and closed) principal ideal (see 1.17).

(2) The function ${}_mf : S \to 2$ defined by ${}_mf^{-1}(0) = Sm$ is a continuous morphism of sup-semilattices.

Moreover, these conditions imply

(3) m is cocompact.

Proof. (1) => (3) is proved analogously to the proof of (3) => (1) in 3.3.

(1)<=>(2) is straightforward by the opposite of 2.3. □

EXAMPLE 3.15. a) Let A be an infinite set and 0,1 two distinct elements which are not contained in A. Let $S = A \cup \{0,1\}$ and define a semilattice multiplication on S by letting 1 be the identity, 0 the zero, and $AA = \{0\}$. Let the topology be the one point compactification topology of the discrete subspace $A \cup \{1\}$. Then S is a $\underline{Z}$-object such that $K(S) = S$, $K_{co}(S) = \{1\}$, but all elements are cocompact. For if $X \subseteq S$ and X is infinite, then X must contain at least two different elements $a_1, a_2 \in A$, and thus $0 = a_1a_2 = \inf\{a_1a_2\} = \inf X$. □

b) Let $T = \{x \in [0,2] \mid x = 1 \pm \frac{1}{n}, n=1,2,\ldots\}$ under min and T' the proto-subsemilattice of all $x \in]0,1[$. Let $S = (T \times \{0\}) \cup (T' \times \{1\}) \cup \{(2,1)\}$. Then S is a $\underline{Z}$-object with the following properties:

(i) S is a sublattice of $[0,2] \times [0,1]$ with component-wise operation.

(ii) S is not a topological lattice (since

$\sup\{(1-\frac{1}{n}, 1), (1+\frac{1}{n}, 0)\} = (2,1)$, but

$\sup\{(1,1), (1,0)\} = (1,1))$

(iii) $K(S) = S \setminus \{(1,0), (1,1)\}$

(iv) For an element $s \in S$ the following statements are equivalent

(1) $s \in K(S)$.

(2) $s \in K_{co}(S)$ (i.e. s is a strong local maximum).

(3) s is compact.

(4) s is cocompact.

(v) $s = \inf\{c \mid c$ compact and $s \leq c\}$ for all $s \in S$ with $s \neq (1,0)$.

In particular, the set of cocompact elements and the set $K_{co}(S)$ are dense, while S is not a topological lattice.

(c) Let $T'' = \{x \in [0,1] \mid x = 0$ or $x = \frac{1}{n}, n=1,2,3,...\}$ under min, and let $S' = S \cup (\{0\} \times T'') \cup (T'' \times \{0\})$ with the multiplication induced from the product $[0,2] \times [0,1]$. Then (i) and (ii) above hold, and the analogue of (iii).

(iii') $K(S') = S \times \{(1,0), (1,1)\}$

Further

(iv') $s \in K(S') \setminus \{(0,0)\}$.

but (v') $K_{co}(S) = \{(2,1)\}$

(vi') $s = \inf\{c \mid c$ cocompact and $s \leq c\}$ for all $s \in S'$ with $s \neq (0,1)$.

Here the set of compact elements is dense, whereas $K_{co}(S)$ is as small as possible, namely, singleton.

Despite the imperfection of the analogy between strong local minima, resp., strong local maxima on one hand and compact, resp. cocompact elements on the other as evidenced in the preceding examples, we have nevertheless the following characterization theorem for those $\underline{Z}$-objects which are, in effect, topological lattices:

THEOREM 3.16. Let S be a $\underline{Z}$-object. Then the following statements are equivalent:

(1) $(x,y) \mapsto \sup\{x,y\} : S \times S \to S$ is continuous,

i.e. S is a topological lattice.

(2) $x \mapsto \sup\{x,a\} : S \to S$ is continuous for all $a \in S$, i.e., S is a semitopological sup-semilattice.

(3) The continuous sup-semilattice characters $S \to 2$ separate.

(4) $s = \inf K_{co}(S) \cap \uparrow s$ for all $s \in S$.

(5) All cocompact elements are strong local maxima and $s = \inf K_{co}(S) \cap \uparrow s$ for all $s \in S$.

These conditions imply

(6) S is cocompactly generated, i.e. $s = \inf\{m \in S \mid m$ is cocompact and $s \leq m\}$,

but the converse may fail. Also (1) - (5) imply

(7) $K_{co}(S)$ is dense.

The converse may fail.

Proof. (1),(2),(3) are equivalent by 1.5, and (1) implies (4) and (5) by 1.12, respectively 3.3. Trivially, (5) implies (4) and (6).

Now assume (4); we will show (3). Let $s \neq t$ in S. We will separate s and t by a continuous sup-semilattice character. Since it suffices to separate st from $\sup\{s,t\}$, we may assume $s < t$. By (4) there is an $m \in K_{co}(S)$ with $s \in Sm$ but $t \notin Sm$. Then ${}_mf(s) = 0$ and ${}_mf(t) = 1$. Clearly (4) => (7) by 1.3.

Example 3.15.a is a cocompactly generated complete lattice, but in the given topology, condition (2) fails; indeed let a_n be a sequence of different elements of A, then $\lim a_n = 0$, hence $\sup\{a_1, \lim a_n\} = \sup\{a_1, 0\} = a_1$ whereas $\lim \sup\{a_1, a_n\} = 1$. Example 3.15.b shows that (7) => (1) may fail. □

Remark. (6) clearly implies (by 1.3).

(6') The set of cocompact elements is dense.

Example 3.15.c shows that (6') may hold in a non-degenerate S, while $K_{co}(S)$ is singleton.

One observes that conditions (4) and (5) are not entirely algebraic; at best they are algebraic in a very technical sense: The given topology is algebraically determined by 3.13, hence the elements of $K_{co}(S)$, which are defined in terms of the given topology, are also determined by the algebraic structure. Example 3.15.a is instructive insofar as it exhibits the difference between the $\underline{Z}$-topology (the given topology), and the $\underline{Z}$-topology determined by the opposite lattice S^{op} by virtue of S being a coalgebraic (= complete, cocompactly generated) lattice (the one point compactification topology of $A \cup \{0\}$). The common refinement is the discrete topology (which is not compact), the common coarsification is the cofinite topology which is quasicompact, but no longer Hausdorff.

For the record, we make the following definition

DEFINITION 3.17. A lattice is called bi-algebraic iff it is algebraic and it is cocompactly generated (i.e., $s = \inf\{m \in S \mid m$ is cocompact and dominates $m\}$ for all s).□ Thus the underlying lattice of a compact zero dimensional topological lattice is bi-algebraic, but not every bi-algebraic lattice carries a compact zero dimensional topology making it into a topological lattice. We will return to this question in the context of distributive lattices in Chapter III.

These results have answered the question: When is an object in $\underline{Z}$ actually a lattice object in the category of compact zero dimensional spaces, i.e. a topological lattice. It does not seem to be possible to obtain such a characterization fully in terms of duality. More precisely we do not have an explicit answer to the following question: Let S be a semilattice; when is its dual $\hat{S}$ a topological lattice? There is a trivial answer to the converse question: Let S be a $\underline{Z}$-object; when is its dual a lattice? We record it for the sake of completeness at this point. We recall a standard definition

DEFINITION 3.18. A lattice L is called arithmetic iff (1) it is algebraic (see Definition 3.5) and (2) K(S) is closed under the operation of taking (finite) infs.

PROPOSITION 3.19. Let S be a $\underline{Z}$-object. Then its dual $\hat{S}$

is a lattice in $\underline{S}$ iff the underlying lattice of S is arithmetic.

Proof. The semilattice $\hat{S}$ is a lattice iff any pair of elements has a sup. In view of 3.7 this is the case iff every pair of elements a,b in $K(S)$ has an inf. Obviously $\inf_{K(S)}\{a,b\} \leq ab$, but by 1.3, 3.6, we must then have $\inf_{K(S)}\{a,b\} = ab$. Hence $K(S)$ is a lattice iff $K(S)$ is closed in S relative to the semilattice multiplication. □

In this first part of Section 3 our main goal has been a description of the dual of a $\underline{Z}$-object S in the form of $K(S)$. In order to have a complete picture of the duality theory, however, we will also discuss morphisms (as we did in Section 2).

PROPOSITION 3.20. Let $f : S \to T$ be a $\underline{Z}$-morphism and define a function $K(f) : K(T) \to K(S)$ by $K(f)(k) = \inf f^{-1}(\uparrow k)$. Then $K(f)$ is an $\underline{S}$-morphism and the diagram

$$\begin{array}{ccc} \hat{T} & \xrightarrow{\hat{f}} & \hat{S} \\ {\scriptstyle \kappa_T}\downarrow & & \downarrow{\scriptstyle \kappa_S} \\ K(T) & \xrightarrow[K(f)]{} & K(S) \end{array}$$

is commutative. In particular, $K : \underline{Z} \to \underline{S}^{op}$ is a functor and $\kappa : \hat{} \to K$ is a natural isomorphism of functors.

Proof. Clearly $f^{-1}(\uparrow k)$ is an open closed filter of S, hence its zero $\inf f^{-1}(\uparrow k)$ is in $K(S)$, thus $K(f)$ is a well-defined function. The remainder will follow from the commutativity of the diagram, since κ is an isomorphism by 3.7. Let $\phi \in \hat{T}$; then $(\kappa_S\hat{f})(\phi) = \kappa_S(\phi \circ f) = \inf(\phi \circ f)^{-1}(1) = \inf f^{-1}[\phi^{-1}(1)]$. Since $\phi^{-1}(1)$ is an open closed filter of T, we have $\phi^{-1}(1) = \uparrow\inf \phi^{-1}(1)$. Hence $(\kappa_S\hat{f})(\phi) = \inf f^{-1}[\uparrow\inf \phi^{-1}(1)] = K(f)(\inf \phi^{-1}(1)) = K(f)(\kappa_T(\phi)) = (K(f)\kappa_T)(\phi)$ which proves the commutativity of the diagram. □

It is useful to have a formula for the inverse of the operation $f \mapsto K(f) : \underline{Z}(S,T) \to \underline{S}(K(T),K(S))$.

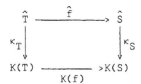

PROPOSITION 3.21. Let $S,T \in \mathrm{ob}\,\underline{Z}$ and suppose that $\psi : K(T) \to K(S)$ is an $\underline{S}$-morphism. Then the unique $f : S \to T$ with $K(f) = \psi$ is given by

$$f(s) = \sup\{k \in K(T) \mid \psi(k) \leq s\} = \sup \psi^{-1}(K(S) \cap Ss)$$
$$= \sup \psi^{-1}(K(Ss)).$$

Proof. By 3.9 we have a natural isomorphism $\gamma_S : S \to \mathfrak{F}(K(S))$, $\gamma_S(s) = K(S) \cap Ss$ with the inverse $F \to \sup F : \mathfrak{F}(K(S)) \to S$. Hence $f = \phi_T{}^{-1}\mathfrak{F}(\psi)\phi_S$ and $f(s) = \sup(\mathfrak{F}(\psi)(K(S) \cap Ss)) = \sup \psi^{-1}(K(S) \cap Ss)$ by 2.4. □

Since we have been able to characterize the objects in $\underline{Z}$ in purely algebraic terms, the question remains, whether the morphism in $\underline{Z}$ can also be characterized in purely lattice theoretical terms.

<u>THEOREM</u> <u>3.22</u>. Let S and T be algebraic lattices, and $f : S \to T$ a morphism of semilattices. Then the following conditions are equivalent:

(1) $\inf f^{-1}(\uparrow k)$ is compact in S and $f(\inf f^{-1}(\uparrow k)) \geqq k$ for each compact $k \in T$

(2) f is continuous relative to the $\underline{Z}$-topologies on S and T (according to 3.10).

(3) (i) $f(\inf X) = \inf f(x)$ for all $X \subseteq S$, and
(ii) $\sup f(Y) = f(\sup Y)$ for all upward directed $Y \subseteq S$.

(4) (i) as in (3) and
(ii') $f(\sup Y) = \sup f(Y)$ for all chains $Y \subseteq S$.

Moreover, if f is a lattice morphism, then these conditions are equivalent to

(5) f preserves arbitrary sups and infs.

Proof. (1) => (2): By (1) $f^{-1}(F)$ is an open closed filter in S for each open closed filter F in T. Since the open closed filters and their complements generate the topologies by 3.13, then f is continuous.

(2) => (3): By 1.3 we have $\inf X = \lim_J \inf F$ where F ranges through the set J of finite subsets of X. Then by (2) we have

$f(\inf X) = f(\lim_J \inf F) = \lim_J f(\inf F) = \lim_J \inf f(F) = \inf f(X)$ by 1.3 again, since every finite subset G of $f(X)$ is of the form $f(F)$ with some $F \in J$. Now let Y be an upward directed set. As before, using 1.3 we have $f(\sup Y) = \lim_J f(\sup F) \geq \lim_J \sup f(F) = \sup f(Y)$, since $\sup f(F) \leq f(\sup F)$. However, since Y is upward directed, for each $F \subseteq J$ there is a $y_F \in Y$ with $y \in y_F$ for all $y \in F$. Then $f(\sup F) \leq f(y_F) \leq \sup f(G)$ for all $G \in J$ with $y_F \in G$. Hence for fixed $F \in J$ we conclude $f(\sup F) \leq \lim_J \sup f(G) = \sup f(Y)$, hence $f(\sup Y) = \lim_J f(\sup F) \leqq \sup f(Y)$. Hence (ii) holds.

(3) => (4) is trivial. Now assume (4), we will show (1). Fix a compact element $k \in T$ and let $s = \inf f^{-1}(\uparrow k)$. Let $Y \subseteq Ss \setminus \{s\}$ be a chain. If $y \in Y$, then $k \nleq f(y)$, for otherwise $k \leq f(y)$, hence $y \in f^{-1}(\uparrow k)$ and so $s \leq y$ in contradiction with $y \in Ss \setminus \{s\}$. Therefore $f(y) \in I(k)$ for all $y \in Y$. Now $f(\sup Y) = \sup f(Y) = \lim_Y f(y) \in I(k)$, since (4 ii') holds and $I(k)$ is closed by 3.3. On the other hand, $f(s) = f(\inf f^{-1}(\uparrow k)) = \inf ff^{-1}(\uparrow k) \geq k$, since (4 i) holds and $ff^{-1}(\uparrow k) \subseteq \uparrow k$. Therefore $\sup Y < s$, so $s \in K(S)$ by 3.3. Thus (1) holds.

Finally, suppose that f is a lattice morphism, i.e. preserves finite sups. Clearly (5) implies (4). Assume (2) and let $X \subseteqq S$. Then $f(\sup X) = f(\lim_J \sup F) = \lim_J f(\sup F) = \lim_J \sup f(F) = \sup f(X)$ where F ranges through the set J of finite subsets of X, and where we have used 1.3, the continuity of f, the preservation of finite sups by f and 1.3 again. □

DEFINITION 3.23. We say that a morphism $f : S \to T$ of semilattices is algebraically continuous iff the following conditions are satisfied:

(a) Whenever $\inf X$ exists in S, then $\inf f(X)$ exists in T and equals $f(\inf X)$.

(b) For every chain $Y \subseteq S$, for which $\sup Y$ exists in S, also $\sup f(Y)$ exists in T and equals $f(\sup Y)$. □

The following is clear:

PROPOSITION 3.24. The class of all semilattices together with all algebraically continuous semilattice morphisms is a subcategory $\underline{CS}$ of $\underline{S}$, containing the full subcategory in $\underline{S}$ of all finite semilattices. □

Much more significantly, however, we have the following conclusive result which characterizes the category $\underline{Z}$ completely in algebraic terms:

THEOREM 3.25. The category $\underline{Z}$ of all compact zero dimensional topological semilattices and continuous semilattice morphisms is isomorphic to the category $\underline{CA} \subseteq \underline{CS}$ of all algebraic lattices and algebraically continuous semilattice morphisms.

Proof. By 3.6, 3.10, and 3.22, the assignment which associates with each $\underline{Z}$-object the underlying algebraic lattice and with each $\underline{Z}$-morphism the underlying algebraically continuous semilattice morphism is a bijective functor $\underline{Z} \to \underline{CA}$. □

Remark. Recall that the isomorphy of two categories is a much stronger property than their equivalence, which for all category theoretical purposes is the really significant concept of equivalence between categories.

In view of the duality theorem I-3.9 we then have

COROLLARY 3.26. The category $\underline{CA}$ is dual to $\underline{S}$. □

This is an instance where we have the duality of a category (viz. $\underline{S}$) with what, on the surface, appears to be a very small subcategory (viz. $\underline{CA}$).

We record the obvious consequence of 3.17 and the subsequent remarks:

COROLLARY 3.27. The full subcategory (in $\underline{Z}$) of topological compact zero dimensional lattices is isomorphic to a proper full subcategory of the category of bialgebraic lattices and algebraically continuous semilattice morphisms. □

HISTORICAL NOTES FOR CHAPTER II.

Compact semilattices have been studied rather extensively, and research in this area is still in flux. In fact, it seems to have been one of the most active of the coherent areas of compact semigroup theory in the last few years. Much of what we collect in Section 1 has been observed from the beginning. Certainly the monotone convergence and the completeness theorems (1.1 through 1.4) appear in the early papers on topological lattices and semilattices. On the other hand, Theorem 1.5 uses a very recent result of Lawson's saying that every compact semitopological semilattice is in fact topological. For our purposes this result is a supplement, and not an essential step in the build-up. Local minima have been used in various contexts, although their crucial role in the case of zero dimensional semilattices has never been so strongly emphasized as in our presentation. On the other hand, in topological monoids, the local maxima or their variants usually play a lesser role; Proposition 1.14 is a first indication that semimaxima at least function significantly in the theory. This result is based on an algebraic theorem which is due to Dilworth and Crawley. [D-3], with a predecessor by Birkhoff and Frink [B-9].

In Section 2 we discuss the equivalence of characters and filters on semilattices. These ideas are, of course, also widely spread throughout the literature where filter spaces have been used as tools for completions of order or topological structures. The strong emphasis on the use of characters is perhaps a bit more recent; in the context of functorial adjunctions and dualities between various categories of partially ordered sets and lattices on the one hand and categories of topological spaces on the other, the correspondence between characters, filters and prime ideals was extensively utilized by Hofmann and Keimel [H-5].

Section 3 is the real bridge between our duality theory between discrete and compact zero dimensional semilattices and algebraic lattice theory. The key results (3.3, 3.6, 3.7, 3.10, 3.22, 3.25, 3.26) are new but they relate closely to a large body of well-established lattice theory. In a sense the link comes from the identification between the local minima in a compact zero dimensional semilattice (where they are an order theoretical and topological concept) with the compact elements of the underlying lattice (where they are a purely lattice theoretical concept). The definition of a compact element in a lattice is due to Nachbin [N-1], although it had a predecessor in the form of Birkhoff's and Frink's join inaccessible element [B-9] (which in algebraic lattices is equivalent to it; as we show in 3.3; an element k is join inaccessible if it satisfies 3.3. (8) with Y upwards directed rather than a chain); this equivalence is also found in Birkhoff [B-8, pp.187-188]. Nachbin introduced the ideal completion of a sup-semilattice and gives necessary and sufficient conditions for a lattice to be the ideal completion of the sub-sup-semilattice of its compact elements. These conditions state that the lattice be algebraic in the now current terminology (Definition 3.5). As we pointed out in a brief digression after 3.14, we retrieve this result from our duality theory, and place it in a new context. Nachbin's theorem is also presented in Birkhoff's book and is credited to Birkhoff and Baker in the formulation given there [B-8, p.187]. Algebraic lattices have developed into a major theory within lattice theory, and they have numerous applications to ideal lattices in rings, lattices of submodules or of ideals in lattices, lattices of subalgebras in universal algebras. Birkhoff and Frink showed that a lattice is isomorphic to the lattice of subalgebras of a suitable abstract algebra A with finitary operations iff it is algebraic. They observed that the lattice of congruences of an abstract algebra with finitary operation is always algebraic, and that a lattice is arithmetic (Definition 3.18) if it is isomorphic to the lattice of ideals of a suitable lattice. Thus the Birkhoff-Frink theorems justify the name "algebraic" for the lattices in question. The literature pertaining to algebraic lattices, compact elements in lattices is vast, and these brief comments can in no way be exhaustive. We conclude by noting that we have not added new contents but rather a new aspect to this body of information, the aspect of compact topological monoid theory and duality.

CHAPTER III. Application of duality to lattice theory

As we have seen in Chapter II the category $\underline{Z}$ of compact zero dimensional semilattices is isomorphic to the category of algebraic lattices and algebraically continuous morphisms. This already opened the door for a connection from the duality between the categories $\underline{Z}$ and $\underline{S}$ to lattice theory in general. In this Chapter we discuss our duality in view of certain facets of lattice theory such as the spectral theory of lattices (i.e. the concept of irreducible and prime elements, their generalizations, modifications, opposites, and the topological spaces associated with lattices via these concepts) or the validity of equations such as the distributive law or Boolean lattices.

It can hardly be expected that strikingly new discoveries about these very classical aspects of lattice theory will come to light. However, it appears that new relations which heretofore were undiscovered emerge through the duality theory which we introduce in the earlier part of this exposition.

Section 1. Primes and duality.

Just as in ring theory, the concepts of prime elements and ideals are of utmost importance in lattice theory. In lattice theory, however, it seems unavoidable to consider a variety of concepts related to the prime property and their opposite concepts. In the following definition we list the concepts systematically, and since our starting point is semilattices (i.e. commutative idempotent monoids) we formulate the concepts in this frame work. Any relation involving $\sup\{a,b\}$ in a semilattice is understood to hold "provided $\sup\{a,b\}$ exists". This is always the case if we are considering lattices.

DEFINITION 1.1. Let S be a semilattice and $x \in S$.

(1) x is meet irreducible (m.i.) iff $x = ab$ implies $x \in \{a,b\}$.

(2) x is join irreducible (j.i.) iff $x = \sup\{a,b\}$ implies $x \in \{a,b\}$.

(3) x is prime iff $x \geq ab$ implies $x \geq a$ or $x \geq b$ (i.e. iff the complement of Sx is a subsemilattice).

(4) x is co-prime iff $x \leq \sup\{a,b\}$ implies $x \leq a$ or $x \leq b$ (i.e. iff $a,b \in I(x)$ implies $\sup\{a,b\} \in I(x)$).

(5) x is completely meet-irreducible (c.m.i.) iff $x = \inf A$ implies $x \in A$.

(6) x is completely join-irreducible (c.j.i.) iff $x = \sup A$ implies $x \in A$.

(7) x is completely prime (c.p.) iff $x \geq \inf A$ implies the existence of an $a \in A$ with $x \geq a$ (i.e. iff the complement of Sx is closed under the formation of arbitrary infs, wherever they exist).

(8) x is completely co-prime iff $x \leq \sup A$ implies the existence of an $a \in A$ with $x \leq a$ (i.e. iff $I(x)$ is closed under arbitrary sups, wherever they exist).

The set of prime elements of S will be called Prime S. □

Inevitably various concepts of distributivity will emerge in our discussions. Here they are.

DEFINITION 1.2. Let S be a semilattice.

(1) S is called weakly distributive if $(\sup\{a,b\})x = \sup\{ax,bx\}$ whenever $\sup\{a,b\}$ exists.

(2) S is called distributive if $\uparrow((\uparrow a \cap \uparrow b)x) = \uparrow ax \cap \uparrow bx$ for all $a,b,x \in S$.

In the following, let S be a lattice:

(3) S is called pre-Brouwerien iff $(\sup A)x = \sup Ax$ whenever $\sup A$ exists.

(4) S is called strongly sup-distributive iff $(\sup A)(\sup B) = \sup\{ab \mid a \in A,\ b \in B\}$ if $\sup A$ and $\sup B$ exist.

(5) S is called Brouwerien if $\max\{s \in S \mid st \leq x\}$ exists for all $t,x \in S$.

REMARK. $\uparrow((\uparrow a \cap \uparrow b)x) \subseteq \uparrow ax \cap \uparrow bx$ holds in any semilattice. Thus (2) is equivalent to

(2') $\uparrow ax \cap \uparrow bx \subseteq \uparrow((\uparrow a \cap \uparrow b)x)$.

PROPOSITION 1.3. In any $\underline{Z}$-object, the conditions (1)-(5) of 1.2 are equivalent.

Proof. If S is a $\underline{Z}$-object, then it is a complete lattice by II-1.4. (3), (4) and (5) are known to be equivalent under these circumstances [B-8]. In a complete lattice, the implications (4) => (3) => (1) <=> (2) are trivial. (In fact, both (1) and (2) reduce to distributivity of a lattice if S is a lattice.)

Suppose now that S is distributive, and let $A,B \subseteq S$. Let I [resp. J] be the set of finite subsets of A [resp. B]. Then $(F,G) \in I \times J$ implies $\sup FG = \sup F \sup G$. Every finite subset of AB is contained in one of the form FG, $(F,G) \in I \times J$. Then $\sup AB = \lim_{I \times J} \sup FG = \lim_{I \times J} (\sup F)(\sup G) = \lim_I \sup F \lim_J \sup G = \sup A \sup B$ by II-1.3. □

REMARK. Every distributive semilattice is clearly weakly distributive. We will give a simple example later which will show that the converse fails (see 1.29). One should note that semilattices which we call weakly distributive have been called distributive in the literature (see e.g. Schein [S-1]). The stronger concept 1.2.(2), however, is more suitable for our purposes in the context of duality. In any event, we will comment on their relationship in more detail later.

Gratzer evolved a concept of distributivity for a semilattice S [G-5] which is as follows:

(2') S is called G-distributive iff for all $x,y,w \in S$ with $w \geq xy$ there is an $a \geq x$ and a $b \geq y$ such that $w = ab$.

We will observe easily, with remarks available in the literature, that S is distributive iff it is G-distributive (Ex. 1.54). For further equivalent formulations see Gaskill [G-1].

SUPPLEMENTS. Conditions (3)-(5) have dual counterparts. In particular, we will draw attention to the following

(4^{op}) S is called strongly inf-distributive iff $\sup\{(\inf A),(\inf B)\} = \inf\{\sup\{a,b\} \mid a \in A, b \in B\}$ if $\inf A$ and $\inf B$ exist.

(6) A complete lattice S is called strongly distributive iff it is strongly sup- and inf-distributive.

The following shows (4) does not imply (4^{op}):

EXAMPLE. Let $T = \{\frac{1}{n} : n = 1,...\} \cup \{0\}$ under min, and $S = T \times 2/(\{0\} \times 2)$. Then $S \in \text{ob } \underline{Z}$, and if $A = \{(1,0)\}$ and $B = (T\backslash\{0\}) \times \{0\}$, then $\inf A = (1,0)$, $\inf B = \overline{0}$, so $\sup\{\inf A, \inf B\} = (1,0)$. However, $\sup\{a,b\} = (1,1)$ for all $a \in A$, $b \in B$; thus $\inf\{\sup\{a,b\} \mid a \in A, b \in B\} = (1,1) \neq \sup\{\inf A, \inf B\}$, whence ($4^{op}$) is violated (as is 1.2(3)). The strongest concept of distributivity is the following:

(7) A complete lattice S is completely distributive if for each family $\{A_x : x \in X\}$, $A_x \subseteq S$, one has $\inf\{\sup A_x : x \in X\} = \sup\{\inf s(X) : s \in \Sigma\}$, where $\Sigma = \{s : X \to \cup A_x \mid s(x) \in A_x \text{ for each } x\}$.

This concept is equivalent to its dual [R-2]. □

PROPOSITION 1.4. Let S be a semilattice and $x \in S$. Then

(1) x is c.p. $\Rightarrow$ x is c.m.i.

$\Downarrow$ $\Downarrow$

x is prime $\Rightarrow$ x is m.i.

(2) If S is distributive and $\sup\{x,s\}$ exists for all $s \in S$, then x is prime iff x is m.i.

(3) If S is dually pre-Browerian, then x is c.p. iff x is c.m.i.

(4) x is completely coprime $\Rightarrow$ x is c.j.i.

$\Downarrow$ $\Downarrow$

x is coprime $\Rightarrow$ x is j.i.

(5) If S is distributive, then x is coprime iff x is j.i.

(6) If S is pre-Browerian, then x is completely

coprime iff x is c.j.i.

(7) If S is an algebraic lattice, then x is compact if x is c.j.i.

Proof. (1)-(6) are standard and left as an exercise.

We prove (7). Let x be c.j.i. and assume $x \leq \sup X$ for some $X \subseteq S$. By I-1.13 and I-1.3 we have $\sup X = \lim_J \sup F$, F ranging through the set J of finite subsets F of X. Then $x = x(\sup X) = x \lim_J \sup F = \lim_J x \sup F = \sup\{x \sup F \mid F \in J\}$. Since x is c.j.i., there is an $F \in J$ with $x = x \sup F$, showing $x \leq \sup F$. □

PROPOSITION 1.5. If S is an algebraic lattice, then every completely prime element is cocompact and every completely coprime element is compact.

Proof. Clear. □

We reformulate a portion of Definition II-1.7.

DEFINITION 1.6. Let S be an algebraic lattice. An element $m \in S$ is called semimaximal iff there is a compact element $k \in S$ such that m is maximal in $I(k)$. (The maximal elements of $I(k)$ are sometimes called the values of k.)

Note that in the light of II-3.6 and II-3.10, 1.6 above and semimaximum (see II-1.7) are indeed equivalent definitions.

PROPOSITION 1.7. In an algebraic lattice S an element is semimaximal iff it is c.m.i.. The subsemilattice T generated by the c.m.i. elements is order dense (i.e. $s = \inf T \cap \uparrow s$ for all $s \in S$). If S is distributive, then s is completely prime iff it is semimaximal and the subsemilattice generated by the prime elements is order dense in S.

Proof. We may assume that S is a $\underline{Z}$-object by II-3.10.

Suppose that m is maximal in $I(k)$ for $k \in K(S)$ and that $m = \inf X$ for some $X \subseteq S$. If $m \notin X$, then $X \subseteq \uparrow k$ by the maximality of m in $I(k)$. Hence $k \leq \inf X = m$ contradicting $m \in I(k)$. Hence m is c.m.i. Conversely, let m be c.m.i., let $m' = \inf\{s \in S \mid m < s\}$.

Then $m < m'$, since m is c.m.i. By 1.12 we find a $k \in K(S)$ with $k \leq m'$ and $k \nleq m$, i.e. $m \in I(k)$. Since $m' \in \uparrow k$, then m is maximal in $I(k)$, i.e. m is semimaximal. The remainder follows from II-1.14 and 1.3 above. □

In particular, in any $\underline{Z}$-object, semimaximal and c.m.i. are the same, and in any distributive $\underline{Z}$-object, semimaximal and c.p. are equivalent properties.

COROLLARY 1.8. If S is a $\underline{Z}$-object, then the subsemilattice generated by the c.m.i. elements of S is dense. If S is distributive, then the subsemilattice generated by the prime elements of S is dense.

Proof. 1.7 and II-1.3. □

With Definition 1.1 it is now easy to define the concepts of prime ideals, meet irreducible ideals, prime filters, meet irreducible filters, etc., if we observe that in a semilattice the collection $\mathcal{J}(S)$ of ideals is a semilattice under $(I,J) \to IJ = I \cap J$ with identity S (parallel to our earlier observation that the set $\mathcal{F}(S)$ of filters is a semilattice under $(F,G) \to F \cap G$, see II-2).

DEFINITION 1.9. Let S be a semilattice. Then an ideal I is prime (meet irreducible) iff I is a prime (meet irreducible) element of the semilattice $\mathcal{J}(S)$. A filter F is prime (m.i.) iff F is a prime (m.i.) element of the semilattice $\mathcal{F}(S)$. The other concepts of Definition 1.1 are carried over to ideals and filters in precisely this fashion. □

REMARK. We note that an ideal P of S is prime according to Definition 1.9 iff $IJ \subseteq P$ implies $I \subseteq P$ or $J \subseteq P$ for all ideals I,J. Hence, if $x,y \notin P$ then $Sx \nsubseteq P$ and $Sy \nsubseteq P$, so $Sxy = (Sx)(Sy) \nsubseteq P$, which implies $xy \notin P$; thus $S\backslash P$ is a subsemilattice. Conversely, if P is prime according to our earlier Definition II-2.2, then $S\backslash P$ is a submonoid; if $I \nsubseteq P$ and $J \nsubseteq P$ we find elements $x \in I\backslash P$ and $y \in J\backslash P$ and so $xy \in IJ\backslash P$, showing that P is prime in the above sense. Therefore Definition 1.9 above is equivalent to Definition II-2.2.

The following observations are straightforward:

PROPOSITION 1.10. In a semilattice S we have the following conclusions: (a) $x \in S$ is prime iff Sx is a prime ideal (b) $x \in S$ is coprime iff $\uparrow x$ is a prime filter. □

PROPOSITION 1.11. Let S be a semilattice and F a filter. Let $I = S \setminus F$ be the complementary ideal. Then the following statements are equivalent:

(1) F is a prime filter

(2) $\uparrow a \cap \uparrow b \subseteq F$ implies $a \in F$ or $b \in F$ for all $a,b \in S$.

(3) I is upwards directed (i.e. $a,b \in I$ implies the existence of a $c \in I$ with $a \leq c$ and $b \leq c$).

Proof. (1) => (2) => (3) is trivial. Assume (3). Let $G,H \in \mathcal{F}(S)$ with $G \not\subseteq F$ and $H \not\subseteq F$. Then there exist $a \in G \setminus F \subseteq I$ and $b \in H \setminus F \subseteq I$, and by (3) we find a $c \in I$ dominating a and b, whence $c \in G \cap H$. Thus $G \cap H \not\subseteq F$, showing that F is a prime filter. □

We now begin to relate these concepts with morphisms.

DEFINITION 1.12. Let $f : S \to T$ be a function between semilattices. We say that f is a sup-morphism iff (1) f is a morphism of semilattices and (2) $\uparrow f(\uparrow a \cap \uparrow b) = \uparrow f(a) \cap \uparrow f(b)$ for all $a,b \in S$. A character $f : S \to 2$ is called a sup-character iff it is a sup-morphism. The corresponding concepts are used in the category $\underline{Z}$. □

PROPOSITION 1.13. If $f : S \to T$ is a sup-morphism and P a prime filter of T, then $f^{-1}(P)$ is a prime filter in S.

Proof. Suppose that $\uparrow a \cap \uparrow b \subseteq f^{-1}(P)$. Then $\uparrow f(a) \cap \uparrow f(b) = f(\uparrow a \cap \uparrow b) \subseteq P$. Since P is prime, $f(a) \in \uparrow f(a) \subseteq P$ or $f(b) \in \uparrow f(b) \subseteq P$, hence $a \in f^{-1}(P)$ or $b \in f^{-1}(P)$, whence $f^{-1}(P)$ is prime by 1.9. □

COROLLARY 1.14. Let f be a character of S. Then the following statements are equivalent:

(1) $f^{-1}(1)$ is a prime filter.

(2) $f^{-1}(0)$ is an upwards directed ideal.

(3) f is a sup-character.

Proof. (1) <=> (2) by 1.11, and (3) => (1) by 1.13 and the fact that $\{1\}$ is a prime filter in 2. Finally assume (1) and take $a,b \in S$. We have $\uparrow f(F) = f(F)$ for any filter F in S. Hence $\uparrow f(\uparrow a \cap \uparrow b) = f(\uparrow a \cap \uparrow b) = \{1\}$ iff $\uparrow a \cap \uparrow b \subseteq f^{-1}(1)$ iff $\uparrow a \subseteq f^{-1}(1)$ or $\uparrow b \subseteq f^{-1}(1)$ by (1), hence $\uparrow f(\uparrow a \cap \uparrow b) = \{1\}$ iff $f(a) = f(b) = 1$ iff $\uparrow f(a) \cap \uparrow f(b) = \{1\}$. Hence f is a sup morphism. □

COROLLARY 1.15. Let $S \in$ ob $\underline{Z}$, $f \in \hat{S}$ a character. Define $k = \min f^{-1}(1) = \kappa_S(f) \in K(S)$. Then the following statements are equivalent:

(1) $f \in$ Prime $\hat{S}$.

(2) f is a sup-character.

(3) $k \in$ Prime $K(S)$.

(4) k is a coprime of S.

(5) k is a complete coprime of S.

In particular, Prime $\hat{S}$ is the set of continuous sup-characters of S.

Proof. By II-2.4, the function $\phi_S : \hat{S} \to \mathcal{F}(S)$, $\phi_S(f) = f^{-1}(1)$ is an isomorphism of semilattices. This, together with 1.10.b and 1.14 shows (1) <=> (2) <=> (4). By II-3.7, (1) <=> (3). (4) => (5): Since k is compact $k \leq \sup X$ implies $k \leq \sup F$ for some finite $F \subseteq X$, and since k is coprime, there is an $x \in F \subseteq X$ with $k \leq x$. Hence (5). But (5) => (4) is trivial. □

REMARK. Notice that 1.15 says in particular that coprimes and complete coprimes are one and the same in any $\underline{Z}$-object (i.e. in any algebraic lattice).

DEFINITION 1.16. Let $f : S \to T$ be a function between semilattices. We say that f is a prime morphism iff (1) f is a morphism of semilattices and (2) $f(p)$ is prime in T whenever p is prime in S. □

Note that every element of 2 is prime, so that every character is a prime character.

COROLLARY 1.17. Let $f : S \to T$ be a sup-morphism between semilattices in $\underline{S}$. Then the dual morphism $\hat{f} : \hat{T} \to \hat{S}$ in

$\underline{Z}$ is a prime morphism.

Proof. By 1.11, the morphism $\mathfrak{F}(f) : \mathfrak{F}(T) \to \mathfrak{F}(S)$ of II-2.4 preserves primes. In view of the natural isomorphism $\phi : \hat{} \to \mathfrak{F}$ of 2.4, the assertion follows. □

We finally observe that the sup-morphisms carry their name justly:

PROPOSITION 1.18 Let $f : S \to T$ be a sup-morphism of semilattices. If $\sup\{a,b\}$ exists in S, then $\sup\{f(a),f(b)\}$ exists in T and equals $f(\sup\{a,b\})$.

Proof. If $c = \sup\{a,b\}$ exists, then $\uparrow a \cap \uparrow b = \uparrow c$. Thus $\uparrow f(c) = \uparrow f(\uparrow a \cap \uparrow b) = \uparrow f(a) \cap \uparrow f(b)$, whence $f(c) = \sup\{f(a),f(b)\}$. □

It is clear that a morphism $f : S \to T$ preserving all existing finite sups will actually satisfy $\uparrow f(\uparrow a \cap \uparrow b) = \uparrow f(a) \cap \uparrow f(b)$ for all $a,b \in S$ for which $\sup\{a,b\}$ exists. Hence we have

COROLLARY 1.19. Let $f : S \to T$ be a semilattice morphism between two lattices. Then f is a sup-morphism iff f is a lattice morphism. In particular, a $\underline{Z}$-morphism is a sup-morphism iff it is a lattice morphism. □

PROPOSITION 1.20. The class of semilattices together with the class of sup-morphisms forms a subcategory of $\underline{S}$ containing the category of lattices with identity and identity preserving lattice morphisms.

Proof. From the definition it is clear that a composition of sup-morphisms is a sup-morphism . The rest follows from 1.19. □

We should point out that it is possible to characterize the precise dual of a prime morphism, although the characterization is purely technical insofar as it amounts just to a translation of the definition:

PROPOSITION 1.21. Let $f : S \to T$ be a morphism in $\underline{S}$. Then the following statements are equivalent:

(1) For each prime filter P of T and each pair of elements $a,b \in S$ the relation $f(\uparrow a \cap \uparrow b) \subseteq P$

implies $f(a) \in P$ or $f(b) \in P$.

(2) $\hat{f} : \hat{T} \to \hat{S}$ is a prime morphism.

Proof. (2) holds iff $f^{-1}(P) = \mathfrak{F}(f)(P)$ is prime in $\mathfrak{F}(S)$ for all primes $P \in \mathfrak{F}(T)$ by II-2.4. For $f^{-1}(P)$ to be prime it is necessary and sufficient by 1.11, that $\uparrow a \cap \uparrow b \subseteq f^{-1}(P)$ always implies that $a \in f^{-1}(P)$ or $b \in f^{-1}(P)$, which is (1). □

PROPOSITION 1.22. If $f : S \to T$ is a morphism of semilattices and the sup-characters of T separate the points (equivalently, the prime filters of T separate the points), then condition (1) of 1.21 implies

(3) f preserves existing finite sups.

Proof. Suppose that $c = \sup\{a,b\}$. Then $f(c)$ is an upper bound for $f(a)$ and $f(b)$. Let $t \in T$ with $t < f(c)$. Then, since the prime filters on T separate the points, there is a prime filter P with $f(c) \in P$ but $t \notin P$. Now $f(\uparrow a \cap \uparrow b) = f(\uparrow c) \subseteq P$, so 1.21(1) implies $f(a) \in P$ or $f(b) \in P$. As $t \notin P$, clearly $f(a) \nleq t$ or $f(b) \nleq t$, so $f(c)$ is indeed the least upper bound of $f(a)$ and $f(b)$. □

REMARK. We will show in 1.28, that the hypothesis on T is equivalent to the distributivity of T.

COROLLARY 1.23. If $f : S \to T$ is a morphism of semilattices for lattices S and T, and if T is distributive, then the following conditions are equivalent:

(1) f is a sup-morphism.

(1') f is a lattice morphism.

(2) $\hat{f} : \hat{T} \to \hat{S}$ is a prime morphism.

Proof. (1') ⟹ (1) is clear from $\uparrow\sup\{x,y\} = \uparrow x \cap \uparrow y$ in a lattice and the definitions. (1) ⟹ (2) was proved in 1.17. Now for a distributive lattice T, the lattice morphisms $T \to 2$ separate the points [G-4]. Hence 1.21 and 1.22 show that (2) ⟹ (1'), thereby finishing the proof. □

DEFINITION 1.24. For a semilattice S we will denote the set of all characters f with $f(\sup\{a,b\})=\sup\{f(a),f(b)\}$

whenever $\sup\{a,b\}$ exists by $\check{S}$. □

Either 1.18 or 1.22 implies

REMARK 1.25. For every semilattice S we have
$$\operatorname{Prime} \hat{S} \subseteq \check{S}.$$

We are now closing in on the characterization of the distributivity of semilattices (see 1.2). The idea of the proof of the following lemma is due to Schein:

LEMMA 1.26. Let S be a semilattice.

a) If S is weakly distributive, then $\check{S}$ separates the points (Schein).

b) If S is distributive, then $\operatorname{Prime} \hat{S}$ separates the points.

Proof. Let $f : S \to 2$ be a character and $P = f^{-1}(1)$. Then $f \in \check{S}$ iff

(A) $\sup\{a,b\} \in P$ implies $a \in P$ or $b \in P$, provided $\sup\{a,b\}$ exists, for all $a,b \in S$.

and $f \in \operatorname{Prim} \hat{S}$ iff f is a sup character (1.15) iff

(B) $\uparrow a \cap \uparrow b \subseteq P$ implies $a \in P$ or $b \in P$, for all $a,b \in S$.

Thus for given $x < y$ we must find a filter P containing y and excluding x which satisfies (A), respectively (B). For later reference, we show a little more generally: If Y is a filter and $x \notin Y$, then there is a filter P with $x \notin P$, $Y \subseteq P$ satisfying (A), respectively (B). By Zorn's Lemma we pick a maximal filter with $x \notin P$ and $Y \subseteq P$. We define $F = \{u \in S \mid pa \leq u$ for some $p \in P\}$ and $G = \{u \in S \mid pb \leq u$ for some $p \in P\}$. Then $F,G \in \mathfrak{F}(S)$ and $P \subseteq F \cap G$. We now assume that $a \notin P$ and $b \notin P$; in case (a) we will show that $\sup\{a,b\}$ (if it exists) is not in P, and in case (b) we will show $\uparrow a \cap \uparrow b \nsubseteq P$. Since $a \in F$ and $b \in G$, we have $P \neq F$ and $P \neq G$, and by the maximality of P we conclude $x \in F$ and $x \in G$, i.e. $x \in F \cap G$. Thus $pa \leq x$ and $qb \leq x$ for some $p,q \in P$. Then $pqa \leq x$ and $pqb \leq x$. In case (a) we assume that $\sup\{a,b\}$ exists and derive $pq(\sup\{a,b\}) = \sup\{pqa,pqb\} \leq x$ with the aid of weak distributivity.

Since $p,q \in P$ but $x \notin P$ we have $\sup\{a,b\} \notin P$. In the case of (b) we conclude $x \in {\uparrow}pqa \cap {\uparrow}pqb = {\uparrow}pq({\uparrow}a \cap {\uparrow}b)$ with the aid of distributivity, which implies ${\uparrow}a \cap {\uparrow}b \nsubseteq P$, since $x \notin P$, but $p,q \in P$. □

LEMMA 1.27. Suppose that S is a semilattice satisfying

$$(*) \qquad s = \inf(\text{Prime } S \cap {\uparrow}s) \quad \text{for all} \quad s \in S.$$

Then Prime $\hat{S}$ separates the points of S.

Proof. As usual it suffices to separate two points $x < y$. By (*) there is a $p \in$ Prime S with $x \leq p$ and with $y \nleq p$, i.e. with $x \in Sp$ and $y \notin Sp$. Now Sp is a prime ideal by 1.10 (b), hence $S \setminus Sp$ is a subsemilattice by the remark following 1.9 and then also a filter. The character $f : S \to 2$ given by $f^{-1}(0) = Sp$ is a sup-character by 1.14. Since $f(x) = 0$ and $f(y) = 1$, the assertion is proved. □

The next theorem is the first main theorem of this section which connects distributivity and the concept of primes via duality. Let us recall that Prime $\hat{S}$ is the set of sup-characters of a semilattice S and that

$$\text{Prime } \hat{S} \subseteq \check{S} \subseteq \hat{S}.$$

<u>THEOREM</u> <u>1.28</u>. The following statements are equivalent for a semilattice S and its character semilattice $T = \hat{S}$:

(1) $\check{S}$ separates the points of S.

(2) The semilattice morphism $ev : S \to 2^{\check{S}}$ given by $ev(s)(f) = f(s)$ is injective and preserves finite sups whenever they exist.

(3) There is a semilattice injection $j : S \to L$ into a distributive lattice which satisfies the following conditions

 (i) j preserves finite sups whenever they exist.

 (ii) The function $f \mapsto fj : \hat{L} \to \hat{S}$ maps the set of lattice characters of L bijectively onto $\check{S}$ (i.e. upon identification of S with its image in L, every $\check{S}$ character is the restriction of precisely one lattice character.

(4) S is isomorphic to an $\cap$-semilattice of subsets of a set such that all existing two element sups are given by $\cup$.

(5) S is weakly distributive.

Secondly, the following statements are equivalent and imply (1) - (5)

(I) The prime filters of S separate filters and points (i.e. if $x \in S$, $x \notin F \in \mathcal{F}(S)$, then there is a $P \in \text{Prime}\,\mathcal{F}(S)$ with $x \notin P$ and $F \subseteq P$).

(II) Every filter in S is the intersection of prime filters.

(III) $t = \inf(\text{Prime } T \cap \uparrow t)$ for all $t \in T$.

(IV) S is distributive.

(V) T is distributive.

Thirdly, the following conditions are equivalent, imply (1) - (5) and are implied by (I) - (V):

(a) Prime $\hat{S}$ separates.

(b) The subsemilattice generated in T by Prime T is dense.

(c) $k = \inf(\text{Prime } T \cap \uparrow k)$ for all $k \in K(T)$.

Proof. (I) => (a) => (1) is clear (since (a) may be rephrased as saying "the prime filters separate points".) The first group: (5) => (1): 1.26. (1) => (2) clearly ev is injective if $\check{S}$ separates. Suppose $\sup\{a,b\}$ exists in S. Then $ev(\sup\{a,b\})(f) = f(\sup\{a,b\}) = \sup\{f(a),f(b)\}$ if $f \in \check{S}$, and $\sup\{f(a),f(b)\} = \sup\{ev(a)(f),ev(b)(f)\} = [\sup\{ev(a),ev(b)\}](f)$. (2) => (3). Let L be the lattice generated in 2^S by ev(S) and let j be the corestriction of ev. (i) is clear. For (ii) we first take a lattice character f of L. Then $fj \in \check{S}$. Since j(S) generates L, then $fj = f'j$ implies $f = f'$, and if $\phi \in \check{S}$, then $f = pr_\phi|L$ with the ϕ-th projection $pr_\phi : 2^{\check{S}} \to 2$, is a lattice character with $fj = pr_\phi\, ev = \phi$. (3) => (5) is obvious, as are (2) => (4) => (5).

The second group: (I) <=> (II) is trivial. (II) <=> (III)

follows from the isomorphism $\phi_S : \hat{S} \to \mathfrak{F}(S)$ of II.2.4. (IV) => (III): 1.7. (III) => (IV): We apply 1.27 to T and with 1.15 and 1.19 obtain that the lattice characters of L separate. Hence T is a distributive lattice by (1) <=> (4) above applied to T. (IV) => (I) If Y is a filter in S and $x \notin Y$, let P be a maximal filter containing Y and excluding x. The proof of 1.26 shows that P is a prime filter. (I) => (V). Let $a,b,x \in S$. We have to show that $\uparrow ax \cap \uparrow bx \subseteq \uparrow(\uparrow a \cap \uparrow b)x$. By (I) this means that every prime filter P containing $\uparrow(\uparrow a \cap \uparrow b)x$ must contain $\uparrow ax \cap \uparrow bx$, i.e. at least one of ax or bx. Thus we must show that for a prime filter P the relations $(\uparrow a \cap \uparrow b)x \subseteq P$ and $ax \notin P$ imply $bx \in P$. Since $1 \in \uparrow a \cap \uparrow b$ we have $x \in P$, hence $a \notin P$ (otherwise $ax \in P$). We claim that $b \in P$, which will furnish the proof of $bx \in P$. For if $b \notin P$, then by 1.11 there is an $s \in S \setminus P$ with $a \leq s$ and $b \leq s$, whence $sx \in (\uparrow a \cap \uparrow b)x \subseteq P$ which is impossible since $sx \leq s \notin P$. The third group. Let T' be the subsemilattice generated in T by Prime T. Since K(T) is dense, and all $\uparrow k$ are open, (b) <=> (c) by II-1.3. In view of the isomorphism $\phi_S : \hat{S} \to \mathfrak{F}(S)$ of II-2.4 which maps elements of K(S) to the principal filters, (a) <=> (c). The following example illustrates that at least (4) does not imply (IV) (i.e. the first group and the second are not equivalent). Thus $(1) \overset{\not\Rightarrow}{\Leftarrow} (a) \Leftarrow (I)$. We know nothing about (a) => (I).

EXAMPLE 1.29. Let S be the following semilattice:

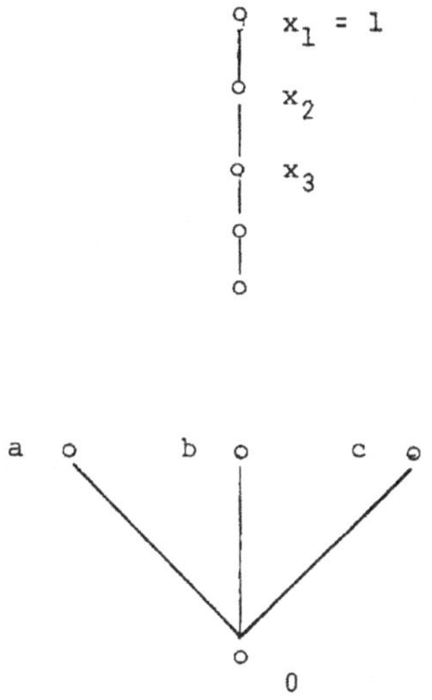

$$1 = x_1 > x_2 > x_3 > \ldots > a,b,c > 0.$$

Then S is weakly distributive, but not distributive. □

We point out the following subtle distinction: By 1.7 every element s in any $\underline{Z}$-object satisfies the following condition

(CMI) $s = \inf\{p \in S \mid s \leq p$ and p is completely meet irreducible$\}$.

In particular, it follows

(MI) $s = \inf\{p \in S \mid s \leq p$ and p is meet irreducible$\}$.

However, it is not always true that

(P) $s = \inf\{p \in S \mid s \leq p$ and p is prime$\}$.

In fact, (P) holds for all $s \in S$ iff the character semilattice S is distributive by Theorem 1.28.

Our goal now is a converse of the preceding theory; we start with a $\underline{Z}$-object S, characterize its prime characters, and find out what it means that the dual of S is primally generated.

PROPOSITION 1.30. Let $S \in \underline{Z}$ be a compact zero-dimensional semilattice, and $f \in \hat{S}$ a character, i.e. a continuous morphism $f : S \to 2$. Then the following statements are equivalent:

(1) $f \in \text{Prime } \hat{S}$.

(2) $f^{-1}(1)$ is a compact open prime filter.

(3) $f^{-1}(0)$ is a compact open lattice ideal.

(4) f is a continuous sup-character.

(5) f is a continuous lattice character

(6) [resp. (6')]. $\min f^{-1}(1)$ is a coprime [resp. a complete coprime] in S.

(7) $\min f^{-1}(1)$ is a prime in $K(S)$.

Proof. (1)<=>(2)<=>(3)<=>(4)<=>(6)<=>(6')<=>(7) by 1.14 and 1.15, in view of the fact that S is a lattice (II-1.4), whence a semilattice ideal is upwards directed iff it is a lattice ideal (i.e. a semilattice ideal which is closed under finite sups). (4)<=>(5) by 1.23. □

DEFINITION 1.31. A semilattice S is primally generated iff it is generated by Prime S, i.e. iff every element is a finite product of primes. For the sake of completeness we will say that a $\underline{Z}$-object S is primally generated if

$$s = \inf(\text{Prime } S \cap \uparrow s) \quad \text{for all} \quad s \in S.$$

Before we produce an analogue to 1.28 we pursue the discussion of II-3.15 ff in which we characterize the topological lattices among the $\underline{Z}$-objects. There we showed that the underlying lattice of a topological lattice in $\underline{Z}$ is bialgebraic (II-3.17). In the presence of distributivity, the converse is true:

Indeed, let S be an algebraic lattice, then every element s is the sup of the c.j.i. elements which it dominates. If, in addition, S is distributive, then every c.j.i. element is completely coprime, hence in particular coprime. By 1.4.(7) or 1.5, every completely coprime element is compact. Hence we have

$$(**) \qquad s = \sup\{k \mid k \in K(S) \cap \downarrow s \text{ and } k \text{ is coprime}\} \quad \text{for all} \quad s \in S.$$

If $k \in K(S)$ and k is coprime, then the character $f_k : S \to 2$ defined by $f_k^{-1}(1) = \uparrow k$ is a Z-continuous lattice character by II-3.3 and 1.15. Let $x < y$ in S. Then by (**) there is a $k \in K(S) \cap Sy$ which is coprime and satisfies $k \notin Sx$, i.e. $x \in I(K)$. Hence $f_k(y) = 1$ and $f_k(x) = 0$. It follows that the $\underline{Z}$-continuous lattice characters separate the points. Hence the morphism $ev : S \longrightarrow 2^{\hat{S}}$, where $ev(s)(f) = f(s)$ and $\hat{S}$ is the set of $\underline{Z}$-continuous lattice characters, is an algebraic and topological embedding of S with its $\underline{Z}$-topology into the distributive topological lattice $2^{\hat{S}}$, so S is a distributive topological lattice. In view of (1)$\Rightarrow$(6) in II-3.16 and 3.17 we have proved the equivalence of (1),(2) and (3) in the following result:

LEMMA 1.32. Let S be a distributive $\underline{Z}$-object. Then the following are equivalent statements:

(1) The underlying lattice of S is bialgebraic (Definition II-3.17).

(2) The continuous lattice characters separate the points.

(3) S is a topological lattice. □

Now we are ready for the counterpart of Theorem 1.28.

<u>THEOREM 1.33</u>. The following statements are equivalent for a compact zero dimensional semilattice S:

(1) The continuous lattice characters separate the points.

(2) Prime $\hat{S}$ separates the points.

(3) The $\underline{Z}$-morphism $ev : S \to 2^{\mathrm{Prime}\ \hat{S}}$ given by $ev(f)(s) = f(s)$ is an embedding (and a lattice morphism).

(4) S is (topologically and algebraically) isomorphic to a sublattice of 2^X for some set X on which the projections $2^X \to 2$ induce precisely the lattice characters of S.

(5) S is a distributive topological lattice.

(6) S is the projective limit of an inverse system of finite distributive lattices and surjective

lattice morphisms (i.e. S is profinite in the category of distributive lattices).

(7) $\hat{S}$ is the union of an upwards directed family of primally generated finite subsemilattices such that the inclusions are prime morphisms.

(8) $\hat{S}$ is primally generated.

(9) K(S) is primally generated.

(10) Every element in S is the sup of the complete coprimes which it dominates.

(11) The underlying lattice of S is a distributive bialgebraic lattice.

(12) The underlying lattice of S is completely distributive (see Supplements after 1.3).

Proof. (1)<=>(2) by 1.30. (2)<=>(3) straightforward. (3)<=>(4) trivial. (4)=>(5) clear since 2^X is a distributive topological lattice. (5)=>(6) was proved by Numakura [N-2]. (6)<=>(7) in view of duality, since a finite semilattice is distributive iff its dual is primally generated by (III)<=>(IV) in 1.28. (7)=>(8) is straightforward. (8)<=>(9) by II-3.7. (9)=>(10) since an element of K(S) is a prime in K(S) iff it is a coprime in S by 1.15. (10)=>(9): Every complete coprime is compact; then apply 1.15 again. (8) =>(2): Straightforward. (5)<=>(11):

Lemma 1.32. Raney shows in [R-1] that (12) is equivalent to

(12') S can be embedded into a product of complete chains such that sups and infs are preserved.

Clearly (4) implies (12'). Conversely, if (12') is satisfied, then S is a compact topological lattice relative to the interval topology. As the dual of $\hat{S}$, it has the compact $\underline{Z}$-topology. Since there is at most one compact semilattice topology on S, by 3.10, these topologies agree. Hence (5) follows. □

If we say briefly that S is a distributive object in $\underline{S}$ iff it satisfies 1.2.2. and S is a distributive object in $\underline{Z}$ iff it satisfies 1.32 (5), then we can formulate

COROLLARY 1.34. Under the duality of $\underline{S}$ and $\underline{Z}$ being primally generated and being a distributive objects are dual properties.

Proof. 1.31, 1.28, 1.32. □

As a complement we observe:

PROPOSITION 1.35. A finite semilattice S is primally generated iff it is distributive.

Proof. Since S is finite, it is a lattice. We recall that $\hat{S}$ is isomorphic to the opposite lattice S^{op} (obtained from S by interchanging the operations, i.e. reversing the order). Now S is distributive iff S^{op} is distributive iff $\hat{S}$ is distributive iff S is primally generated (by 1.32). □

The following example shows that the infinite analogue may fail:

EXAMPLE 1.36. Let $S = (\mathbb{N} \times \mathbb{N}) \cup \{\infty\}$, $\mathbb{N} = \{1,2,3,...\}$ with ∞ as identity and otherwise componentwise lattice operations. Then S is a distributive lattice, but it has no primes other than 1, hence fails to be primally generated.

We now have a complete characterization of distributivity and generation by primes expressed in the following main theorem of this section

THEOREM 1.37. Let S be a semilattice and T its dual in $\underline{Z}$. Then the following statements are equivalent:

(1) S is a distributive semilattice.

(2) T is a distributive lattice.

(3) T is primally generated (1.31).

(4) T is pre-Brouwerian (1.2)

(5) T is Brouwerian (1.2)

(6) T is dually pre-Brouwerian (i.e. the opposite lattice T^{op} is pre-Brouwerian).

Furthermore the following conditions are equivalent:

(I) S is primally generated.

(II) T is a $\underline{Z}$-topological distributive lattice.

(III) T is a bialgebraic distributive lattice.

(IV) T is a completely distributive algebraic

lattice.

Condition (I) implies condition (1) but the converse fails.

Proof. (2),(4) and (6) are equivalent by 1.3. (1) <=> (2) by 1.28. (I)<=>(II)<=>(III) by 1.33. (I)=>(2) trivial. (1)$\not\Rightarrow$(I) is illustrated by Example 1.36. The implication (III)<=>(IV) was observed in 1.33. □

For the morphisms we record:

PROPOSITION 1.38. Let $f : S \to T$ be a $\underline{Z}$-morphism and $\hat{f} : \hat{T} \to \hat{S}$ its dual. We consider the following statements

(1) f is a lattice morphism.

(2) $\hat{f}$ is a prime morphism.

(3) f is a prime morphism.

(4) $\hat{f}$ is a sup-morphism.

(D) T is distributive.

(d) $\hat{T}$ is distributive.

Then (1)=>(2), (4)=>(3), (D)<=>(d) and [(2) and (d)] implies (1).

Proof. By 1.17 we have (1)=>(2) and (4)=>(3), and (d)<=>(D) was established in 1.37. By 1.23, (D) implies the equivalence of (1) and (2). □

The results of II-3, notably II-3.24, II-3.27 and 1.35, 1.36 above now immediately yield the following duality theorems:

<u>THEOREM</u> <u>1.39</u>. The category of distributive semilattices and prime morphisms is dual to the category of Brouwerian algebraic lattices and lattice morphisms preserving arbitrary sups and infs. □

<u>THEOREM</u> <u>1.40</u>. The category <u>DBA</u> of distributive bialgebraic lattices and all lattice morphisms preserving arbitrary sups and infs is isomorphic to the category of topological compact zero dimensional distributive lattices and continuous lattice morphisms, and is dual to the category of $\underline{S}_p$ of primally generated semilattices and prime morphisms.

Proof. 1.33, 1.38 and II-3.27. □

The category of Theorem 1.39 has been also identified

in different ways by Hofmann and Keimel [H-5].

DEFINITION 1.41. Let us call a topological space X spectral if it satisfies the following conditions: (i) X is T_0. (ii) Every irreducible subset is a singleton closure (where a set is irreducible iff it is closed and not contained in the union of two proper closed subsets. (iii) X has a basic of quasi-compact open sets (i.e every open set is the union of the quasi-compact open subsets which it contains). The category of spectral spaces contains these spaces and all continuous maps.

The following is due to Hofmann and Keimel [H-5,pp.50, 51].

PROPOSITION 1.42. The category of spectral spaces is dual to the category of algebraic Brouwerian lattices and lattice morphisms which preserve arbitrary sups. □

One direction of this duality is given by the functor which associates with a Brouwerian algebraic lattice L the space Prime L of primes with the hull-kernel topology, the other by the functor which associates with a spectral space X the lattice O(X) of open sets and with a continuous map $f : X \to Y$ the morphism O(f) with $O(f)(V) = f^{-1}(V)$. The question now arises in our context: When will O(f) preserve arbitrary infs?

LEMMA 1.43. If $f : X \to Y$ is a continuous map of topological spaces, then O(f) preserves infs iff
(0) For each collection $\mathcal{U}$ of open sets in Y we have Interior $f^{-1}(\cap\mathcal{U}) = f^{-1}(\text{Interior} \cap \mathcal{U})$.

This condition is satisfied if f is open, i.e. the image of every open set is open, and if Y is a T_1-space, then O(f) preserves arbitrary infs iff f is open.

Proof. Firstly, condition (0) is obviously equivalent to the preservation of arbitrary infs. Now f is open iff
(0') For each subset A we have Interior $f^{-1}A = f^{-1}(\text{Interior } A)$.
(Indeed $\supseteq$ is clear; if f is open then f(Interior $f^{-1}A$) is open and contained in $ff^{-1}A \subseteq A$, whence the inclusion

$\subseteq$ follows. Conversely, if (0') is satisfied, and U is open in X, set $A = f(U)$, then $U \subseteq f^{-1}f(A) = f^{-1}A$ hence $U \subseteq \text{Interior } f^{-1}A = f^{-1}(\text{Interior } A) = f^{-1}(\text{Interior } f(U))$, whence $f(U) \subseteq \text{Interior } f(U) \subseteq f(U)$, so $f(U)$ is open.) Finally, if Y is T_1, then for any set A we have $A = \cap\{X\backslash\{b\} \mid b \in X\backslash A\}$ and $X\backslash\{b\}$ is open for each b, so that (0) implies (0'). □

DEFINITION 1.44. We say that a function $f : X \to Y$ between topological spaces is semi-open iff (0) of 1.43 holds.

Thus a function f between spaces is continuous and semi-open iff $O(f) : O(Y) \to O(X)$ preserves arbitrary sups and infs (i.e., is an algebraically continuous lattice morphism).

EXAMPLE 1.45. Let $X = \{0,1\}$ with the discrete topology and Y the T_0-space $\{0,1\}$ with topology $\{Y,\{1\},\phi\}$. Then the identity map $f : X \to Y$ is continuous and semi-open, but not open. □

Now we can put 1.39 and 1.42 together and conclude:

COROLLARY 1.46. The categories of distributive semilattices and prime morphisms on one hand and spectral spaces and semi-open continuous maps are equivalent categories. □

EXERCISE. Give the functors defining the equivalence of 1.46 explicitly, using the functors giving the duality in 1.39 and the duality in 1.42.

In the following remarks we wish to characterize the category $\underline{S}_p$ of primally generated semilattices and prime morphisms in terms of the category $\underline{PO}$ of partially ordered spaces. We recall from I-1.9 that the objects of $\underline{PO}$ are partially ordered sets with maximal element and the morphisms are order preserving maps respecting the greatest elements. It was observed that the forgetful functor $\underline{S} \to \underline{PO}$ has a left adjoint $\Sigma : \underline{PO} \to \underline{S}$ which associates with a poset X the set ΣX of all non-empty finite subsets $F \subseteq X$ such that $(F \times F) \cap \text{graph} \le \, = \emptyset$ (such sets are said to be unrelated) such that the operation $FG = \min(F \cup G) =$ set of minima of $F \cup G$ gives the semilattice multiplication; moreover, Σ associates with a $\underline{PO}$-morphism $f : X \to Y$ the $\underline{S}$-morphism $\Sigma f : \Sigma X \to \Sigma Y$ given by

$$(\Sigma f)(F) = \min f(F).$$

We first wish to observe that Σ maps $\underline{PO}$ into the category of primally generated semilattices and prime maps:

LEMMA 1.47. Let $X \in$ ob $\underline{PO}$. Then Prime $\Sigma(X) = \{\{x\} | x \in X\} \cong X$. In particular, ΣX is primally generated.

Proof. Let $P \in$ Prim $\Sigma(X)$, and let $x \in P$. We claim $P = \{x\}$. If not, set $F = P \setminus \{x\}$; then $\{x\}, F \in \Sigma X$ and $\{x\}F = \min(\{x\} \cup F) = \min P = P$. Since P is prime, $\{x\} \le P$ or $F \le P$. In the first case, $\{x\} = \{x\}P = \min(\{x\} \cup P) = \min P = P$; in the second, $F = FP = P$ which is incompatible with $\{x\} = P \setminus F$. This proves the claim. Conversely, we prove that $\{x\}$ is prime in ΣX for all $x \in X$. Suppose $FG \le \{x\}$, i.e. $\min F \cup G \cup \{x\} = \min F \cup G$. Assume $F \not\le \{x\}$, i.e. $\min F \cup \{x\} \ne \min F = F$, i.e. $\min F \cup \{x\} = F \cup \{x\}$. Since x is a minimal element of $F \cup G$ we have $x \in G$. Thus $G\{x\} = \min G \cup \{x\} = \min G = G$, i.e. $G \le \{x\}$. Since ΣX is clearly generated by $\{\{x\} \mid x \in X\}$, the Lemma is proved. □

LEMMA 1.48. If $f : X \to Y$ is a $\underline{PO}$-morphism, then Σf is a prime morphism.

Proof. $\Sigma(f)(\{x\}) = \{f(x)\}$; the assertion then follows from 1.45. □

DEFINITION 1.49. If the category of primally generated semilattices and prime morphisms is denoted by $\underline{S}_p$, then the corestriction of the functor $\Sigma : \underline{PO} \to \underline{S}$ to the smaller codomain $\underline{S}_p$ is well-defined by 1.43 and 1.44 and will also be denoted by Σ.

We record the following result due to Horn and Kimura [H-11].

PROPOSITION 1.50. Let $S \in$ ob $\underline{S}_p$. Then $\Sigma(\text{Prime } S) \cong S$.

Proof. The universal property of the left adjoint yields a unique morphism $f : \Sigma(\text{Prime } S) \to S$ given by $f(F) = \wedge F$; since S is primally generated, f is surjective. We show that f is injective: Let $F, G \in \Sigma(\text{Prime } S)$ with

$\wedge F = \wedge G$. Let $p \in G$. Then $\wedge F = \wedge G = \wedge(F \cup G) \leq p \wedge F$, so there is $q \in F$ with $q \leq p$ as p is prime. Similarly, there is $r \in G$ with $r \leq q$, so $r \leq q \leq p$. But G consists of unrelated elements of S, so $r = p$, whence $q = p$. Therefore $p \in F$. Thus $G \subseteq F$, and the reverse containment follows dually. □

It is clear that the assignment of the PO-object Prime S to an $\underline{S}_p$-object S extends to a functor Prime : $\underline{S}_p \to \underline{PO}$, and our preceding results show:

THEOREM 1.51. The categories $\underline{S}_p$ and $\underline{PO}$ are equivalent under the pair of functors Prime : $\underline{S}_p \to \underline{PO}$ and $\Sigma : \underline{PO} \to \underline{S}_p$. □

COROLLARY 1.52. The category $\underline{PO}$ of partially ordered sets with greatest element and the category $\underline{DBA}$ of distributive (Brouwerian) bialgebraic lattices and lattice morphisms preserving arbitrary infs and sups are dual.

Proof. Immediate from 1.51, 1.40. □

We supplement the preceding discussion by describing the duality between $\underline{PO}$ and $\underline{DBA}$ a bit more explicitly. Let X be a poset with maximal element. Let $S = \Sigma X$; we may then identify X with Prime S. If $f \in \hat{S}$ is a character of S, then $f|X : X \to 2$ is a PO-character; conversely, by the universal property of ΣX, every PO-character $\phi : X \to 2$ extends to a unique S-character $f : S \to 2$ with $\phi = f|X$. Hence $\hat{S}$ may be identified with the collection $\underline{PO}(X,2)$ of all PO-characters equipped with the pointwise algebraic and topological structure inherited from 2^X. Thus, the functor Ch $= \hat{} \circ \Sigma : \underline{PO} \to \underline{Z}$ is given in a standard fashion [H-5] by the formation of a character object. On the other hand, let $T \in$ ob $\underline{Z}$. The elements f of Prime $\hat{T}$ are those continuous characters for which min $f^{-1}(1)$ is a (complete) coprime. The partially ordered space Prime $\hat{T}$ is therefore order-anti-isomorphic to the poset of all (complete) coprimes of T with the partial order induced by the semilattice partial order of T. The functor Prime $\circ \hat{}: \underline{Z} \to \underline{PO}$ is therefore naturally isomorphic to the functor Coprime^{op} which associates with a

$\underline{Z}$-object the $\underline{PO}$-object of the set Coprime of (complete) coprimes with the opposite order (making 0 the maximal element). (We leave it as an exercise to formulate the definition of Coprime (f) for a $\underline{Z}$-morphism f.) From the preceding results we know that the functor Coprime $\circ$ Ch is naturally isomorphic to the identity functor of $\underline{PO}$ (i.e. $X \cong$ Coprime(Ch(X)), naturally). Similarly, if $T \in$ ob DBA, then T = Ch(Coprime (T)) , naturally. The functor Coprime : $\underline{S} \to \underline{PO}^{op}$ is left adjoint to Ch : $\underline{PO}^{op} \to \underline{Z}$, and Ch $\circ$ Coprime : $\underline{Z} \to \underline{DBA}$ is a left reflector. In particular, $\underline{DBA}$ is a left reflective subcategory of $\underline{Z}$. At this point, it need no longer be particularly emphasized that the functor Ch : $\underline{PO} \to \underline{DBA}$ can be expressed in terms of filters on posets rather than in terms of characters, where the function $f \mapsto f^{-1}(1)$: Ch X $\to \mathfrak{F}(X)$ sets up an isomorphism between the lattice of $\underline{PO}$-characters of $X \in$ ob $\underline{PO}$ and the lattice of filters on X.

In particular, since $\mathfrak{F}(X)$ is a complete ring of sets (i.e. a collection of subsets of X which is closed under arbitrary unions and intersections) then we may derive

COROLLARY 1.53. Let S be a distributive algebraic lattice. Then the following are equivalent assertions

(A) S is bialgebraic (i.e. S satisfies the dozen equivalent conditions of 1.33).

(B) S is isomorphic to a complete ring of sets.

Proof. (A) => (B) was just observed.

(B) => (A): If (B) is satisfied, then S is a completely distributive algebraic lattice, hence satisfies (A) by 1.37. □

EXERCISES.

EXERCISE 1.54. Show that for a semilattice S the following statements are equivalent:

(1) S is distributive (1.2.(2)).

(2) S is G-distributive (Remark following 1.2).

(3) $\hat{S}$ is distributive.

(Indication of Proof. (1)<=>(3) by 1.28. Green [G-6] and

Gratzer [G-4, p.118] show that (2) holds iff $\mathcal{J}(S)$ is distributive. But $\hat{S} = \mathcal{J}(S)$ by II-2.4. □

Section 2. Duality and Boolean lattices.

Among the distributive lattices, the Boolean ones are the most classical. In this area the first duality theorem was proved which involved a category of lattices and topological spaces: The well-known Stone duality theorem.

DEFINITION 2.1. A semilattice S is called pseudocomplemented iff it has a zero 0 and for each $s \in S$

$$s^{\perp} = \max\{t \in S \mid st = 0\}$$

exists. The function $\perp$ is called a pseudocomplementation. If a pseudocomplementation is an involution, i.e. satisfies

$$s^{\perp\perp} = s \quad \text{for all} \quad s \in S,$$

then it is called a complementation. A semilattice with a complementation is called a Boolean semilattice. A lattice with a complementation is called a Boolean lattice. □

LEMMA 2.2. A Boolean semilattice is a Boolean lattice.
Proof. If $s,t \in S$, observe that $\sup\{s,t\} = (s^{\perp}t^{\perp})^{\perp}$. [Compare H-5, p.73 ff.] □

DEFINITION 2.3. A Boolean object in $\underline{S}$ [resp. $\underline{Z}$] is a Boolean lattice in $\underline{S}$ [resp. a Boolean lattice in $\underline{Z}$]. A morphism $f : S \to T$ of Boolean objects (Boolean morphism, in short) is a (semilattice) morphism between Boolean objects in $\underline{S}$, resp. $\underline{Z}$ which in addition satisfies

$$f(a^{\perp}) = f(a)^{\perp}.$$

Note that every such morphism is automatically a lattice morphism. A character is called a Boolean character if it is a Boolean morphism. The set of Boolean characters of a Boolean lattice S is denoted $S_{\perp} \subseteq \check{S} \subseteq \hat{S}$. Note that 2 is a Boolean object in $\underline{S}$ and $\underline{Z}$ with $0^{\perp} = 1$, $1^{\perp} = 0$. □

LEMMA 2.4. Any product of Boolean objects is a Boolean

object (under componentwise operations) in $\underline{S}$ or $\underline{Z}$. □
For Boolean objects, one has a well-known equivalent description in terms of algebras:

PROPOSITION 2.5. Let S be a Boolean object in $\underline{S}$ [resp. $\underline{Z}$]. For $a,b \in S$ define

$$a + b = (a \vee b)(ab)^{\perp}.$$

Then S together with $+$ and the given multiplication becomes a Boolean algebra [resp. topological Boolean algebra] (i.e. ring with identity and characteristic 2). Moreover, $a^{\perp} = 1 - a = 1 + a$. A map $S \to 2$ is a Boolean character iff it is a [continuous] morphism of Boolean algebras. In particular, $S_{\perp} = \text{Spec } S$ where $\text{Spec } S$ is the set of all [continuous] algebra maps $S \to 2$. □

Proof. Straightforward. □

LEMMA 2.6. Let S be a Boolean object in $\underline{S}$. If $k \in S$, then the function $m : S \to Sk \times Sk^{\perp}$ defined by $m(s) = (sk, sk^{\perp})$ is a Boolean isomorphism in $\underline{S}$ [resp. $\underline{Z}$] with an inverse given by $m^{-1}(a,b) = a + b$.

Proof. Straightforward verification. □

PROPOSITION 2.7. Let T be a Boolean object in $\underline{Z}$, $k \in K(T)$, and $g \in \hat{T}$ such that $k = \min g^{-1}(1)$. Then the following are equivalent:

(1) $g \in T_{\perp} = \text{Spec } T$.

(2) $I(k) = Tk^{\perp}$.

(3) $Tk = \{0,k\}$, with $k \neq 0$, i.e. k is an atom in T.

(4) $Tk \cap K(T) = \{0,k\}$.

Proof. (1)$\Rightarrow$(2). We have $s \in I(k)$ iff $g(s) = 0 = 1^{\perp}$ iff $g(s^{\perp}) = 1$ iff $k \le s^{\perp}$ iff $s \le k^{\perp}$.

(2)$\Rightarrow$(3): $Tk = (Tk \cap I(k)) \cup \{k\} = (Tk \cap Tk^{\perp}) \cup \{k\} = \{0\} \cup \{k\}$.

(3)$\Rightarrow$(4) trivial.

(4)$\Rightarrow$(1). Since $Tk \cap K(T)$ is dense in Tk by II-1.3, we have $Tk = \{0,k\}$. Then by 2.6 above we have a commutative diagram of morphisms

$$\begin{array}{ccc} T & \xrightarrow{\ m\ } & Tk \times \{0,k\} \\ {\scriptstyle g}\downarrow & & \downarrow{\scriptstyle pr_2} \\ 2 & \xrightarrow[\cong]{} & \{0,k\} \end{array}$$

Since m and pr_2 are Boolean morphisms, so is g. □

DEFINITION 2.8. For $s \in S$ with S a semilattice, we denote with $A(s)$ the set $\{x \in S \mid Sx \cap \uparrow s = \{s,x\}$ and $s < x\}$ of atoms relative to s. □

LEMMA 2.9. In a Boolean object T in $\underline{Z}$ the set $\{k \in K(T) \mid \kappa_T^{-1}(k) \in \text{Spec } T\} = A(0) \cap K(T)$ is free.

Proof. Let $k_1,\ldots,k_n$ be atoms, and let $k = \sup\{k_1,\ldots,k_n\}$. By 2.6, T is isomorphic to $Tk \times Tk^{\perp}$. In order to prove the freeness of the family $k_1,\ldots,k_n$ in T, it therefore suffices to prove it in Tk. But Tk is a Boolean lattice which is generated by $k_1,\ldots,k_n$ and is, therefore, finite. By Stone duality, [H-5] a finite Boolean lattice S is a free semilattice, since it is of the form $2^X = {}^X2$ for the finite set $X = \text{Spec } S$. The atoms in 2^X are precisely the elements $(s_x)_{x \in X}$ with $s_y = 1$ for one $x = y$ and $s_y = 0$ for $x \neq y$. This is a free set. Hence $k_1,\ldots,k_n$ forms a free family. □

LEMMA 2.10. In a Boolean object T in $\underline{Z}$ the sup semilattice $K(T)$ is free.

Proof. By 2.9 it suffices to show that the atoms A generate $K(T)$. Let $t = \sup A$. If $c < 1$, $c \in K(T)$, then there is a $g \in \text{Spec } T$ with $g(c) = 0$ and $g(1) = 1$, since the Boolean characters separate the points by 2.5 and Stone duality. Let $k = \kappa_S(g) = \min g^{-1}(1)$. Then k is an atom by 2.9, and $k \not\leq c$, so $t \not\leq c$. Hence $t = \vee A = 1$. Moreover, for c as above, from $c < \sup A$ we conclude that there is a finite set F of atoms with $c \leq \vee F$. Thus c is an element of the (free) subsemilattice generated by F and is, therefore, contained in the subsemilattice generated by A. □

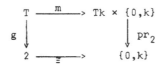

THEOREM 2.11. Let S be a semilattice in $\underline{S}$ and T its dual in $\underline{Z}$. Then the following statements are equivalent:

(1) S is free (over $\underline{\text{Set}}$)

(2) T is Boolean (in $\underline{T}$).

In particular, T is a Boolean object in $\underline{Z}$ iff it is isomorphic to 2^X for some set X. Thus Boolean objects in $\underline{Z}$ are always topological lattices.

Proof. (1)=>(2): If S is free, then $S = {}^X 2$ for some set X, since the free functor preserves coproducts. Thus $\hat{S} = 2^X$ by duality. (2)=>(1): If T is Boolean, then $K(T)$ is free by 2.9. Since $K(T) \cong \hat{S}$ by II-3.7, the assertion follows. □

LEMMA 2.12. Let X be a compact zero dimensional space and $T = \Gamma X$ (the $\underline{Z}$-semilattice of compact subsets of X under $\cup$). Then $U \in K(\Gamma X)$ iff U is an open closed subspace of X.

Proof. By II-3.3, U is compact in ΓX iff U is a local minimum in ΓX iff $\uparrow U = \{A \in \Gamma X \mid A \subseteq U\}$ is open in ΓX. If U is open in X, then $\uparrow U = \{A \in \Gamma X \mid A \subseteq U\}$ is open. Suppose the latter condition is satisfied. The map $x \to \{x\} : X \to \Gamma X$ is an embedding; the inverse image of $\uparrow U$ under this map is U. Hence U is open. □

Recall that Γ is the left adjoint of the grounding $\underline{Z} \to \underline{\text{ZComp}}$ of $\underline{Z}$ into the category of compact zero dimensional spaces. Hence ΓX is $\underline{\text{ZComp}}$-free.

LEMMA 2.13. Let $T \in \underline{Z}$ be $\underline{\text{ZComp}}$-free. Then $S = \hat{T}$ is Boolean (in $\underline{S}$).

Proof. The set $K(\Gamma X)$ of open closed subsets of X is a Boolean lattice. □

LEMMA 2.14. Let S be a Boolean object in S. Let $X = \text{Spec } S$ be the Stone dual of S with the compact zero dimensional topology induced from $\text{Spec } S \subseteq 2^S$. Then $\hat{S} \cong \Gamma X$.

Proof. By duality we must show that $S \cong \widehat{\Gamma X}$. But $\widehat{\Gamma X} = K(\Gamma X)$ = Boolean lattice of open closed sets of X. By the Stone duality, this Boolean lattice has spectrum X.

Hence, by Stone duality, $S \cong \Gamma\hat{X}$. □

LEMMA 2.15. In 2.14, the isomorphism $\alpha_S : \hat{S} \to \Gamma X$ maps the Boolean characters to the singletons.

Proof. The explicit definition of α is $\alpha_S(\phi) = \{\psi \in \operatorname{Spec} S \mid \psi \geq \phi\}$. This implies the assertion. □

THEOREM 2.16. Let S be a semilattice in $\underline{S}$ and T its dual in $\underline{Z}$. Then the following statements are equivalent:

(1) S is Boolean (in $\underline{S}$).

(2) T is ZComp-free. □

Thus 2.10 and 2.16 together roughly say that Boolean and free are dual properties. Let us pursue this observation for morphisms also.

DEFINITION 2.17. A co-atom in a semilattice S is an element $a \in S$ with $\uparrow a = \{a,1\}$ and $a \neq 1$. A morphism $f : S_1 \to S_2$ is co-atomic if $f(a)$ is a co-atom whenever a is a co-atom.

LEMMA 2.18. i) The co-atoms of a free semilattice are precisely the generators $x \in X$.

ii) Let $f : F(X) \to F(Y)$ be a semilattice morphism. Then f is co-atomic iff it is set-induced, i.e. iff there is a function $\phi : X \to Y$ with $f = F(\phi)$.

Proof. Straightforward. □

PROPOSITION 2.19. Let $f : T_1 \to T_2$ be a $\underline{Z}$-morphism between Boolean $\underline{Z}$-semilattices and let $\hat{f} : \hat{T}_2 \to \hat{T}_1$ be its dual. Then $\hat{f}$ is co-atomic (i.e. set induced) iff f is Boolean.

Proof. Since the Boolean characters of T_2 separate the points, f is Boolean iff ϕf is Boolean for all Boolean characters ϕ of T_2. But ϕ is Boolean iff $\phi \in \hat{T}_2$ is a co-atom by 2.7. Since $\phi f = \hat{f}(\phi)$, the assertion follows. □

PROPOSITION 2.20. Let $f : S_1 \to S_2$ be an $\underline{S}$-morphism between two Boolean lattices and let $g : \Gamma(X_2) \to \Gamma(X_1)$, $X_j = \operatorname{Spec} S_j$ be its dual. Then g is space induced (i.e. there is a continuous $\psi : X_2 \to X_1$ with $g = \Gamma(\psi)$)

iff f is Boolean.

Proof. Again f is Boolean iff ϕf is Boolean for all Boolean characters ϕ. Since $\phi f = \hat{f}(\phi)$, and under the isomorphism $\alpha_S : \hat{S} \to \Gamma(\text{Spec } S)$ the Boolean characters map precisely onto the singletons (2.15), we deduce that f is Boolean iff $g(\{x\}) = \{\psi(x)\}$ for a continuous map ψ obtained from g upon restriction to the singletons.

Note: The singletons are the co-atoms of $\Gamma(X)$, hence the space induced Z-morphisms are exactly the co-atomic ones.

Before we summarize our results we make the following simple observations: The functor Σ of 1.49 above will map a partially ordered set of the form $X \cup 1$ with a set X of unrelated elements below 1 onto the free semilattice $F_S X$, and conversely, the functor Prime, when applied to a free semilattice $F_S X$ will single out the partially ordered subset $X \cup 1$. The functor $X \mapsto X \cup 1$ maps the category Set of sets and functions faithfully into PO; if we compose this functor with Σ we clearly obtain an equivalence from the category Set to the category of free semilattices and set induced maps. On the other hand, Prime $F_Z X = X \cup \{\emptyset\}$ for any compact zero dimensional space X (where we recall that $\emptyset$ is the isolated identity of $F_Z X$), and thus, in an analogous fashion, the functor $X \mapsto F_Z X$ defines an equivalence between the category ZComp of compact zero dimensional spaces and continuous maps and the category of ZComp-free objects and space induced maps.

A tabulation of our main results in this section now looks as follows:

THEOREM 2.21. a) The category of free semilattices and co-atomic (i.e. set-induced) maps is equivalent to the category Set and dual to the category of Boolean Z-objects and Boolean Z-maps (which in turn is isomorphic to the category of compact zero-dimensional topological Boolean algebras and continuous algebra morphisms) and also isomorphic to the category of all algebraic Boolean lattices and Boolean maps preserving arbitrary sups.

b) The category of Boolean lattices and Boolean maps is dual to the category of ZComp-free objects and co-atomic (i.e. space-induced) maps, and hence also to the category ZComp.

The dualities are induced by the standard duality between S and Z. □

Clearly b) is just amplification of the classical Stone duality (which we have utilized in our proofs).

We note various consequences of our results. Since every Boolean Z-object turns out to be of the form 2^X, we may record the following

PROPOSITION 2.22. Let T be a Boolean algebra and a compact zero dimensional space such that all functions $x \mapsto ax : T \to T$, $a \in T$ are continuous. Then $T = 2^X$ for some set X. □

A slightly more algebraic characterization of the Boolean lattice 2^X is implicit in 2.21:

PROPOSITION 2.23. Any algebraic and Boolean lattice is of the form 2^X for some set X. □

EXERCISES.

EXERCISE 2.24. Let B be a complete Boolean lattice. Then the following statements are equivalent:

(1) B is completely distributive.

(2) [resp. (2^{op})] B is atomic [resp. co-atomic].

(3) [resp. (3^{op})] B is algebraic (i.e. B is a Z-object with a suitable topology) [resp. B^{op} is algebraic].

(4) B is bialgebraic.

(5) B is a compact topological lattice.

(6) $B = 2^X$ for some set X.

(7) B is a complete field of sets.

Indication of Proof. For those parts not proved in the preceding part of the Section, refer to the references Birkhoff [B-8], Papert-Strauss [P-1], Raney [R-2], Sikorski [S-6], Tarski [T-1].

Section 3. Projectives and injectives in $\underline{S}$ and $\underline{Z}$.

In this section we describe more fully the injectives and projectives in $\underline{S}$ and $\underline{Z}$; we recall that some general facts have been assembled in I-4. Much of the more detailed material in this section was first established by Horn and Kimura [H-11], but our proofs are largely independent unless specific reference is made.

PROPOSITION 3.1. A semilattice S is injective in $\underline{S}$ iff it is a retract (direct factor) of a complete Boolean lattice.

Proof. By I-4.17 and duality, S is injective iff $\hat{S}$ is a retract of $\Gamma(E)$ for some extremally disconnected space E, iff S is a coretract (direct factor) of some $\Gamma(E)^{\wedge}$. But by Section 2 above (notably 2.12 ff.), $\Gamma(E)^{\wedge}$ is a Boolean lattice L with $\operatorname{Spec} L \cong E$, and conversely, if L is a Boolean lattice then $\hat{L} = \Gamma(E)$ with $E = \operatorname{Spec} L$. Finally, since a Boolean lattice L is complete iff $\operatorname{Spec} L$ is extremally disconnected, the assertion follows. □

PROPOSITION 3.2. Every semilattice S in $\underline{S}$ [resp. $\underline{Z}$] can be embedded in an injective object which is, in addition, a complete Boolean object in $\underline{S}$ (resp. $\underline{Z}$).

Proof. If $S \in \operatorname{ob} \underline{S}$, there is a surjection $F_{\underline{Z}}|\hat{S}|_d \cong \Gamma(\beta|\hat{S}|_d) \to \hat{S}$ (the back adjunction of $F_{\underline{Z}}$), whose dual is the required injection.

If $S \in \operatorname{ob} \underline{Z}$, there is a surjection $F_{\underline{S}}(|\hat{S}|) \to \hat{S}$ whose dual satisfies the requirements. □

LEMMA 3.3. Let $P \in \operatorname{ob} \underline{PO}$ (see I-1.9) be such that

(fin) $\uparrow p$ is finite for all $p \in P$.

Let $|P|$ be the underlying set and $F|P|$ the free semilattice over $|P|$ (considered as the $\cup$-semilattice of finite subsets of P). Then the surjection $f : F|P| \to \Sigma P$ defined by $f(A) = \min A$ (see I-1.9) is a retraction in $\underline{S}$.

Proof. Let $X \in \Sigma P$ be an unrelated finite subset of P. Then $f^{-1}(X) = \{Y \subseteq P \mid Y$ finite and $X \subseteq Y \subseteq \cup\{\uparrow p \mid p \in X\}\}$ By (fin) the set $\cup f^{-1}(X)$ is finite, hence is an element

$\tilde{f}(X) \in F|P|$. Note that $\tilde{f}(X) = \inf f^{-1}(X)$ in $F|P|$. Now $\tilde{f} : \Sigma P \to F|P|$ is a PO-morphism. In particular, the function $\phi : P \to F|P|$ given by $\phi(p) = \tilde{f}(\{p\})$ is a PO-morphism. By the universal property of ΣP, there is then a unique S-morphism $\phi' : \Sigma P \to F|P|$ with $\phi'(\{p\}) = \tilde{f}(\{p\})$. Then $f(\phi'(\{p\})) = f\tilde{f}(\{p\}) = \{p\}$ since $f\tilde{f} = 1_{\Sigma P}$. But this property characterizes $1_P : \Sigma P \to \Sigma P$, hence $f\phi' = 1_P$. This shows that f has a right inverse. □

HISTORICAL REMARK. The application of the device ΣP in this general context is due to Horn and Kimura, although the details of their approach are a bit different.

We are now ready for the characterization theorem for projectives in S and injectives in Z:

THEOREM 3.4. Let $S \in$ ob S and $T \in$ ob Z its dual. Then the following statements are equivalent:

(1) S is projective.

(2) S is a retract (direct factor) of a free semilattice FX.

(3) S is a distributive lattice such that $\uparrow s$ is finite for all $s \in S$.

(4) S is primally generated and $\uparrow p \cap \text{Prime } P$ is finite for all $p \in \text{Prime } S$.

(5) $S = \Sigma P$ for some poset $P \in$ PO such that $\uparrow p$ is finite in P for all $p \in P$.

(I) T is injective.

(II) T is a topological direct factor of 2^X for some set X.

(III) T is an (arithmetic) Brouwerian lattice in Z such that Tk is finite for all $k \in K(T)$.

(IV) T is a distributive topological lattice and Tk is finite for all $k \in K(T)$.

Proof. (1)<=>(2): I-4.13. (2)=>(3)=>(4) are trivial, since the primes of FX are precisely the elements $\{x\}$, $x \in X$. (4)=>(5) by 1.50 with $P = \text{Prime } S$. (5)=>(2) by 3.3. The conclusions (I)<=>(1), (II)<=>(2), (III)<=>(3) and (IV)<=>(4) follow from duality. (See in particular 1.33.)

Notice that "arithmetic" in (III) is a consequence of the finiteness of Tk for all $k \in K(T)$.

REMARK. One should notice that this theorem characterizes in particular the injectives in the category of algebraic lattices and algebraically continuous morphisms as the (arithmetic) Brouwerian lattices such that all principal ideals generated by compact elements are finite.
The following lemma is again due to Horn and Kimura [H-11].

LEMMA 3.5. Let S be a complete Brouwerian lattice, $f : A \to S$ an $\underline{S}$-morphism. If A is a subsemilattice of B, then the function $f' : B \to S$ defined by $f'(b) = \sup\{f(a) \mid a \in A \text{ and } a \leq b\} = \sup f(A \cap Bb)$ is an $\underline{S}$-morphism which extends f. Thus S is injective.

Proof. Clearly f' is order preserving, whence $f'(b_1b_2) \leq f'(b_1)f'(b_2)$. Since S is Brouwerian, we have $f'(b_1)f'(b_2) = \sup f(A \cap Bb_1)\sup f(A \cap Bb_2) = \sup f(A \cap Bb_1)f(A \cap B_2) = \sup f[(A \cap B_1)(A \cap Bb_2)] \leq \sup f(A \cap Bb_1b_2) = f'(b_1b_2)$ (since $a_1 \leq b_1$ and $a_2 \leq b_2$ implies $a_1a_2 \leq b_1b_2$). Thus f' is an $\underline{S}$-morphism with $f'(a) = \sup f(A \cap Ba) = \sup f(Aa) = f(a)$ for $a \in A$. □

The next result is the characterization theorem for injectives in $\underline{S}$ and projectives in $\underline{Z}$:

THEOREM 3.6. Let $S \in \text{ob}\,\underline{S}$ and $T \in \text{ob}\,\underline{Z}$ its dual. Then the following statements are equivalent:

(1) S is injective.

(2) S is a direct factor (in $\underline{S}$) of 2^X for some set X.

(3) S is a direct factor (in $\underline{S}$) of a complete Boolean lattice.

(4) S is a complete Brouwerian lattice.

(I) T is projective.

(II) T is a direct (topological) factor in $\underline{Z}$ of a free $\underline{Z}$-object (i.e. some object $\Gamma(\beta X)$).

(III) T is a direct (topological) factor (in $\underline{Z}$) of some $\Gamma(E)$ for an extremally disconnected compact space E.

(IV) T is distributive and K(T) is complete.

Proof. (1)<=>(2) by I-4.13 (b). (2)=>(3) clear. (3)=>(4) follows from the fact that a retract of a complete lattice is complete. (4)=>(1) by Lemma 3.5. That these conditions are equivalent to any one of (I),(II) and (III) again follows from duality (see Section 1 above). (4)=>(IV). Since S is Brouwerian complete, it is, in particular, distributive. Hence T is distributive by 1.28. Since $\hat{S} \cong K(T)$ by II-3.7, then K(T) is complete. (IV)=>(4): Since K(T) is complete, then S is a complete lattice by II-3.7 again. If T is distributive, then it is co-pre-Brouwerian (see 1.2). Hence K(T) is pre-Brouwerian, and since K(T) is complete, it is Brouwerian. Hence S is Brouwerian. □

REMARK. The preceding Theorem shows in particular, that an object in the category of algebraic lattices and algebraically continuous maps is projective iff it is a distributive arithmetic lattice such that the lattice of compact elements is complete.

HISTORICAL NOTES FOR CHAPTER III.

As far as lattice theory is concerned, again in this Chapter, as in the preceding one, we are touching familiar subjects. However, some results and some proofs of known results are probably new. Section 1 is a sequence of variations on the theme of distributivity. This concept poses a bit of a problem for semilattices, since it is by no means clear how one should define distributivity in this case; nor is it clear which one of numerous possibilities is the most suitable one for a given purpose, once such possibilities are found. Two concepts are natural, the weaker one given in 1.2.(1) was used by Schein [S-1] in showing that every semilattice with this property may be considered as a semiring of sets. The stronger one given in 1.2.(2) is based on the idea that in a semilattice one should replace the non-existing sup of two elements a and b by the always existing "virtual sup" $\uparrow a \cap \uparrow b$ in the formulation of distributivity (recall that semilattices always have identities according to our convention). It turns out that this type of distributivity is precisely the one which is useful in the context of duality. In alternative forms, this type of distributibity was introduced by Grätzer [G-4] and discussed by Green [G-6]. For further developments, see also Gaskill [G-1]. Many of the preparatory propositions such as 1.7 are known and difficult to track to their beginnings; 1.7 is credited to Dilworth and Crawley [D-3], but has its forerunners in Birkhoff and Frink [B-9]. The concept of a sup-morphism is an idea which parallels our concept of distributivity insofar as it is a morphism between semilattices which comes as close to a lattice morphism as it conceivably can by respecting "virtual sups". This idea and the special case of a sup-character is new, as are some simple results which relate this concept to duality, such as 1.15 where sup-characters are characterized as the primes in the character semilattice. The core results of the first section are Theorems 1.28, 1.33, 1.37, 1.39, 1.40, 1.51, 1.52 and they are new in general, with portions covering familiar ground. In Theorem 1.28 the different concepts of distributivity of semilattices are characterized in terms of character theory, and in Theorem 1.33 the same is done for $\underline{Z}$-objects, i.e. for algebraic lattices. The bi-partition we are making here is a little loose: In Theorem 1.28 it is in effect shown that a semilattice is distributive iff its character semilattice (an algebraic lattice) is distributive; thus, 1.28 is at the same time a characterization of distributivity both for semilattices and algebraic lattices. Theorem 1.33 on the other hand describes precisely when a $\underline{Z}$-object is a distributive topological lattice, which is, in a sense, the appropriate concept of distributivity for the category $\underline{Z}$ even though it is stronger than just plain distributivity of the underlying lattice. It appears that this stronger concept of distributivity for algebraic lattices is particularly symmetric in the light of the equivalent conditions 1.33 (11) and (12) which says that the underlying lattice has an algebraic (distributive) lattice as opposite lattice, respectively, be an algebraic completely distributive lattice. Theorem 1.37 is just a summary of preceding results which contrapose distributivity and generation by primes. These results use a theorem proved by Raney [R-1] on completely distributive lattices. Theorem 1.46 relates the present duality theory to duality theories between lattices and topological spaces discussed by Hofmann and Keimel [H-5]. More in this direction could and should be done. Theorem 1.51, which states that the categories of posets with 1 and the category of primally generated semilattices with 1 are equivalent, uses a device due to Horn and Kimura [H-11]. This result entails a duality between the category of posets with 1 and the category of distributive bialgebraic (or, completely distributive algebraic) lattices.

Corollary 1.53 rectifies a Theorem of Raney's [R-2, Theorem 2]. Raney's theorem stated that a complete lattice is a complete ring of sets iff every element is a sup of completely join irreducible elements. The later condition is certainly necessary, but not in general sufficient. By 1.7, every algebraic lattice has the property that it is complete and that every element is the inf of completely meet irreducible elements, but it need not even be distributive, and even if it is distributive, it need not be bialgebraic by the results of Section 1.

In Section 2 we inspected the relation of the duality to Boolean algebra and Boolean lattices. Boolean theory is perhaps the most classical of all of the various areas in lattice theory, and one would certainly not expect to make any substantially new contributions here. Again, the point of our discussion is to illustrate that the duality also applies to this situation. The main result is Theorem 2.21 which roughly says that "Boolean" and "free" are dual properties, in a sense made precise. This result in itself is presumably new. However, its ramifications and corollaries over-lap with very classical results. E.g. there is Tarski's classical result [T-1] that a complete and completely distributive topological lattice is of the form 2^X, recently complemented by Dona Papert Strauss [P-1] who showed that a compact topological complete Boolean lattice is completely distributive, hence, by Tarski's theorem, of the form 2^X.

This relates to a theorem of Katetov's [K-1] which says that a Boolean lattice is a compact Hausdorff space in its interval topology iff it is isomorphic with 2^X for some set X. It may be that the characterization 2.23 of 2^X has not been formulated precisely in this way. We describe which of the duals of S-objects are, in addition, Boolean lattices in terms of compact monoids as ZComp-free objects whose structure we describe earlier in Chapter I. An algebraic characterization of these objects has been given by Nachbin [N-1] as arithmetic distributive lattices in which every prime ($\neq$ 1) is a co-atom.

Section 3 is more or less an elaboration of the results of Horn and Kimura [H-11] supplemented with results emerging from the duality and presented with different proofs in various places. Further material on injectives and projectives in related categories may be found in the papers authored and coauthored by Balbes [B-3,4,5,6], see also Grätzer [G-4, pp.143-147].

CHAPTER IV. Applications of Duality to the Structure Theory of Compact Zero Dimensional Semilattices

In Chapter III we proposed applications of the duality theory to lattice theory. In the present Chapter we investigate consequences for the compact monoid structure theory of $\underline{Z}$-objects.

Section 1. Cardinality invariants.

There are several cardinality invariants of a topological space which, in a suitable sense, characterize the "size" of the space:

DEFINITION 1.1. Let X be a topological space. We define cardinals $w(X)$ and $d(X)$ as follows:

$w(X) = \min\{a \mid$ there is a basis for the topology of X with cardinal $a\}$.

$d(X) = \min\{a \mid$ there is a dense subset in X of cardinal $a\}$.

The cardinal $w(X)$ is called the weight of X and $d(X)$ the separability number.

REMARK. Clearly $w(X) = \min\{a \mid$ there is a subbasis for the topology of X with cardinal $a\}$, and $d(X) \leqq w(X)$. A space satisfies the second axiom of countability iff $w(X) \leqq \aleph_0$ and it is separable iff $d(X) \leqq \aleph_0$.

This section serves to describe algebraically the weight and the separability number of a $\underline{Z}$-object. We begin with a purely topological Lemma:

LEMMA 1.2. Let X be a locally compact T_2 space and Y a space. Let $CO(X,Y)$ denote the space of all continuous functions $f : X \to Y$ with the compact open topology. Then

$$w(CO(X,Y)) \leq \max\{w(X),w(Y)\},$$

provided this maximum is infinite.

Proof. The proof is left as an exercise. $\square$

THEOREM 1.3. Let S be a compact zero dimensional semilattice. Then $w(S) = \text{card } \hat{S} = \text{card } K(S)$.

Proof. If S is finite, the result is clear. Therefore, suppose $\text{card } (S) \geq \aleph_o$. We have algebraic and topological embeddings $S \to CO(\hat{S},2)$ and $\hat{S} \to CO(S,2)$ by duality. Since the weight of a closed subspace does not exceed the weight of the whole space, using 1.2, we conclude $w(S) \leq w(CO(\hat{S},2)) \leq \max\{w(\hat{S}),2\} \leq w(S)$ since S is infinite, and similarly, $w(\hat{S}) \leq w(S)$. Thus $w(S) = w(\hat{S})$ if S is infinite. □

REMARK. This portion of the theory is parallel to that of compact abelian groups G, where $w(G) = \text{card } \hat{G}$. The proof is essentially the same.

PROPOSITION 1.4. Let S be a compact zero dimensional semilattice. Then $w(S) \leq 2^{d(S)}$.

Proof. Let T be an $\underline{S}$-object and D a dense set in $\hat{T}$. Then D separates the points of T: Indeed let $d(t_1) = d(t_2)$ for all $d \in D$, $t_j \in T$. By duality we may consider t_j as a continuous character on the $\underline{Z}$-object $\hat{T}$; under this identification, t_1 and t_2 agree on the dense subspace D of $\hat{T}$, hence by continuity they must agree everywhere, i.e. $t_1 = t_2$. Thus T can be injected into 2^D by the evaluation map ev (given by $ev(t)(d) = d(t)$). We apply this to $T = \hat{S}$ and a dense subset D of $S \cong \hat{T}$ of cardinality $d(S)$. Using 1.3, we deduce $w(S) = \text{card } (\hat{S}) \leq \text{card } 2^D = 2^{d(S)}$. □

LEMMA 1.5. Let X be a set of infinite cardinal a and let $S = {}^X 2$ the copower in $\underline{Z}$ (i.e. $S \cong F_{\underline{Z}}(X) \cong \Gamma(\beta X) \cong \alpha(F_{\underline{S}}X)))$. Then $w(S) = 2^{d(S)} = 2^a$.

Proof. We have $w(S) = \text{card } \hat{S} = \text{card } 2^X = 2^a$ by 1.3 and I-4.3. Since $F_{\underline{S}}(X)$ is dense in $\alpha(F_{\underline{S}}(X))$ we have $d(S) \leq \text{card } F_{\underline{S}}(X) = \text{card } X$ (since the underlying set of $F_{\underline{S}}(X)$ is the set of the finite subsets of X). Thus $2^{d(S)} \leq 2^X = w(S)$. The reverse inequality follows from 1.4. □

DEFINITION 1.6. We say that a semilattice S is dominated iff for every cardinal a with card $S \le 2^a$ there is a family $\{S_j \mid j \in J\}$ of semilattices S_j such that

(a) card $J \le a$, card $S_j \leqq a$ for all a.

(b) There is an injective $\underline{S}$-morphism $S \to \Pi S_j$.

In Example 1.10 we will observe that an inversely well-ordered uncountable chain with cardinality at most continuum is not dominated. Clearly every free semilattice is dominated.

LEMMA 1.7. Let $S \in \underline{Z}$ be infinite and suppose that its dual is dominated. Then for every cardinal a with $w(S) \le 2^a$ there is a dense subsemilattice D in S with card $D \le a$.

Proof. By 1.3 we have card $\hat{S} = w(S) \le 2^a$. Since S is dominated, there is an injection $\hat{S} \to \Pi S_j$ as in 1.6. Hence, by duality, there is a surjection

$$\coprod_j \hat{S}_j \to S.$$

By I-4.4.a. the domain of this surmorphism contains a subset which is bijectively equivalent to $\coprod_{\underline{S}} S_j$, whose cardinality is $\le a$ card $J = a^2 = a$. Let D be the subsemilattice generated by the image of this set. □

DEFINITION 1.8. For a cardinal a we write $\log a = \min\{b \mid a \le 2^b\}$.

THEOREM 1.9. Suppose that S is a $\underline{Z}$-object. Then

$$w(S) \geqq d(S) \geqq \log w(S) = \log \text{card } \hat{S},$$

and for an infinite S with dominated dual, equality holds. The estimates are the best possible, and inequality does occur in both places.

Proof. From 1.4 we conclude $\log \text{card } \hat{S} \le d(S)$ and from 1.7 $d(S) \le \log \text{card } \hat{S}$, if S is infinite and dominated. See 1.5 and 1.10 for the last assertion. □

REMARK. Compact semilattices deviate here from compact groups: For a compact group G one has $d(G) = \log w(G)$. [I-2].

EXAMPLE 1.10. Let S be the chain of all ordinals $1,2,\ldots,\Omega$ up to the first uncountable one. Then $w(S) = \text{card } K(S) = \text{card}[1,\Omega[= \aleph_1$ and $d(S) = \aleph_1$. But $\log \aleph_1 = \aleph_0$. In particular, the chain $\hat{S} = [1,\Omega[$ with the inverse order is not dominated.

Section 2. Chains and Stability

In the general theory of compact monoids the totally ordered compact submonoids play an important role [H-7], and in the theory of semilattices, so do the totally ordered subsemilattices, called chains. In this section we report on the applications of duality to the utilization of chains in the structure theory of $\underline{Z}$-objects. These results are based on recent research of the authors [H-6], and we shall refer to that paper for almost all of the proofs of the results in this section. I.e., we regard this section mainly as a report on the work in this area, and our goal is to provide an outline of the results in [H-6] and to give several illuminating examples, rather than a detailed catalogue of the proofs involved.

The principal question to which we shall address ourselves is that of the dimensional stability of $\underline{Z}$-objects: When does a compact zero dimensional semilattice have a quotient semilattice with positive topological dimension? The following axiom demonstrates that the particular dimension function DIM on the category of locally compact spaces which one wishes to employ is irrelevant.

AXIOM 0. If DIM is a dimension function of the category of locally compact spaces, then a compact space has DIM dimension zero if and only if its topology has a basis of open closed sets. (See [H-8].)

In particular, a compact space which is not totally disconnected has positive DIM dimension. We therefore make the following definition.

DEFINITION 2.1. A compact zero dimensional (= totally disconnected) topological semigroup is <u>stable</u> if and only if each homomorphic image is zero dimensional (= totally disconnected); otherwise it is <u>instable</u>.

Lawson has shown that a compact zero dimensional semilattice S is instable iff S has the unit interval I under min multiplication as a quotient. While this reduction shows that we may confine our considerations to the chain quotients of a $\underline{Z}$-object S, it also brings the main difficulty of this investigation into sharp focus. Namely, I is not a $\underline{Z}$-object; in fact, each character of I is constant, and so the duality theory which we have built up for $\underline{Z}$ cannot be brought directly to bear on the problem. As we shall soon see, however, those chain quotients of a $\underline{Z}$-object S which are also $\underline{Z}$-objects, in particular, the existence of quotients which are isomorphic to C, the Cantor chain, forms a sufficient criterion to test the dimensional instability of S. Thus, since we wish to apply our duality theory to this situation, our first task is to characterize the dual of a $\underline{Z}$-object which is a chain, especially the dual $\hat{C}$ of the Cantor chain C.

First note that, by II-3.7, for a $\underline{Z}$-object S, $\hat{S}$ is isomorphic to $(K(S),\vee)$, and so, if S is a chain, then so is $\hat{S}$. Dually, if $S \in \underline{S}$ is a chain, the set $\mathfrak{F}(S)$ of filters is totally ordered by inclusion. Hence by II-2.4, $\hat{S}$ is totally ordered. Furthermore, if $S \in \underline{Z}$ is a well-ordered set under $st = \max\{s,t\}$, then $K(S) = S$, so $\hat{S}$ is just S under $st = \min\{s,t\}$.

To obtain our characterization of $\hat{C}$, we first recall that, if $\mathbb{Q}$ is the set of rational numbers, then a classical result of Cantor's states that any countable order dense chain is order isomorphic to one of the following chains:

i) $[0,1] \cap \mathbb{Q}$ iii) $]0,1] \cap \mathbb{Q}$

ii) $[0,1[\cap \mathbb{Q}$ iv) $]0,1[\cap \mathbb{Q}$

Clearly we may replace $\mathbb{Q}$ by the set $\frac{1}{2^\infty}\mathbb{Z}$ of dyadic rationals, and we denote the min chain $]0,1] \cap \frac{1}{2^\infty}\mathbb{Z}$ by Q. The result we have been pursuing is then obtained in the following

PROPOSITION 2.2. Let S be a compact zero dimensional semilattice. Then the following statements are equivalent:

(1) $S \approx C$.

(2) S is a perfect (separable) metric chain.

(3) $K(S)$ is an order dense countable chain not containing 1.

(4) $\hat{S}$ is an order dense countable chain in $\underline{S}$ without zero.

(5) $\hat{S} \approx Q$.

Furthermore, every countable chain in $\underline{S}$ is a quotient of Q, and every compact metric zero dimensional chain in $\underline{Z}$ can be injected into C. □

A particular consequence of this characterization is that all metric perfect chains in $\underline{Z}$ are isomorphic to C. However, there are perfect separable chains in $\underline{Z}$ which do not contain any metric subchains:

EXAMPLE 2.3. Consider the product $I \times 2$ in the lexicographic order, and let $S = \{(t,n) \in I \times 2 \mid t = 0 \Rightarrow n = 1$ and $t = 1 \Rightarrow n = 0\}$ with the order topology (which is also the topology induced from $I \times 2$). Then S is a perfect separable chain, since $Q \times \{0\}$ is dense in S. However, S contains no non-degenerate subchain which is metric in the induced topology.

Combining Proposition 2.2 with our duality theorem for $\underline{Z}$ yields the following:

PROPOSITION 2.4. If $S \in \underline{Z}$, then the following are equivalent:

1) There is a surmorphism $f : S \to C$.

2) There is an injection $Q \hookrightarrow \hat{S}$.

3) $K(S)$ contains an order-dense countable chain containing 0.

4) S contains a compact chain S' such that, for some order dense countable chain L we have

$$L \subseteq S' \cap K(S) \subseteq S' = \overline{L}. \quad □$$

Note that if S satisfies one of the above conditions, then there is a surmorphism $f : S \to C$, so $\rho f : S \to I$ is a surmorphism, whence S is instable.

COROLLARY 2.5. Under the conditions of 2.4, if S' is

metrizable, then S' contains a chain $S'' \simeq C$, but the identity of S'' need not be 1.

That the assumption of metrizability of S' in 2.5 is necessary is shown in the following

EXAMPLE 2.6. Let $S = \{(t,n) \in C \times 2 :\ t$ is isolated from above in $C \Rightarrow n = 0\}$, $f : S \to C$ be projection, and $L = K(C) \times 2$. Then $S' = S$ but S does not contain a Cantor chain. (Here again, we consider $C \times 2$ in the lexicographic order.)

As a consequence of Corollary 2.5, if $S \in \underline{Z}$ is metrizable and $f : S \to C$ is a surmorphism, then there is a Cantor chain $C \subseteq S$. However, the existence of an injection $C \subseteq S$ is not sufficient for the instability of S, as the following example illustrates.

EXAMPLE 2.7. Let $T \in \underline{S}$ be given by $T = \{(x,y) \in I \times I : y = 1, \frac{1}{2}, \frac{1}{3}, \ldots,\ x \in \{m/2^n : n = 1, \ldots, 2^{y^{-1}}\}\}$ with operations induced from $I \times I$. Then $p : T \to Q$, the projection onto the first coordinate, is a surjection, but T contains no order dense chain. Hence, by duality, if $S = \hat{T}$, then there is an injection $C \hookrightarrow S$, but there is no surjection $S \to C$. It will follow from our main theorems that S is stable. Note that the breadth of S and T is 2.

We now come to the core results of this section, which are a sequence of theorems which show that, in particular, $S \in \underline{Z}$ is instable if and only if there is a surmorphism $f : S \to C$. We denote the set of all surmorphisms of the compact monoids $A \to B$ by $\mathrm{Sur}(A,B)$. $\underline{M}$ is the category of compact monoids and continuous identity-preserving homomorphisms.

THEOREM 2.8. The First Modification Theorem. Let $S \in \underline{Z}$ and J an arbitrary compact chain with identity. Let $f \in \mathrm{Sur}(S,J)$ and let $P \subseteq J$ be any perfect zero dimensional subchain with $1 \in P$. Define $\underline{F}, \overline{F} : J \to P$ by

$$\underline{F}(r) = \sup\{p \in P : p \leq r\}$$

$$\overline{F}(r) = \inf\{p \in P : r \leq p\}.$$

Then there is a $g \in \mathrm{Sur}(S,P)$ with

$$\underline{F}(f(s)) \leq g(s) \leq \overline{F}(f(s)) \quad \text{for each} \quad s \in S.$$

In particular, the equalizer $E = \{s \in S : f(s) = g(s)\}$ is precisely $f^{-1}(P)$. □

Note that the theorem applies in particular to $J = I$ and $P = C$. Furthermore, with this choice for J and P, we can indeed choose P in I with $0 \in P$ and such that $|\underline{F}(r) - \overline{F}(r)| < \varepsilon$ for each $r \in I$ for any prescribed $\varepsilon > 0$. In this sense, we can approximate $f : S \to I$ up to an arbitrarily small error. However, there is another way to express this result as we see in the following

<u>THEOREM</u> <u>2.9</u>. <u>The Second Modification Theorem</u>. Let $S \in \underline{Z}$ and $\rho : C \to I$ the fixed standard Cantor morphism. Then for each $f \in \mathrm{Sur}(S,T)$ and each $\varepsilon > 0$ there are

(a) a $g \in \mathrm{Sur}(S,C)$, and

(b) an $h \in \mathrm{Sur}(I,I)$ with

i) $hf = \rho g$, i.e. the following diagram commutes

$$(\delta) \qquad \begin{array}{ccc} S & \xrightarrow{\;f\;} & I \\ {\scriptstyle g}\downarrow & & \downarrow{\scriptstyle h} \\ C & \xrightarrow[\;\rho\;]{} & I \end{array}$$

ii) $r \leq h(r) < r + \varepsilon$ for each $r \in I$.

Moreover, (δ) is a push-out diagram in $\underline{M}$ (i.e. if $\phi : C \to I$ and $\psi : I \to I$ are $\underline{M}$-morphisms with $\phi g = \psi f$, then there is a unique $\pi : I \to I$ with $\phi = \pi\rho$ and $\psi = \pi h$). □

The following result is an obvious corollary of Proposition 2.4 and either of Theorems 2.8 and 2.9; we regard this as a full characterization of the dimensionally stable objects in $\underline{Z}$.

<u>THEOREM</u> <u>2.10</u>. If $S \in \underline{Z}$, the following are equivalent:

1) S is instable.

2) There is at least one surmorphism $S \to I$.

3) There is at least one surmorphism $S \to C$.

4) $K(S)$ contains an order dense countably infinite

chain containing 0.

5) There is an injection $Q \hookrightarrow \hat{S}$. $\square$

Theorem 2.9 has the following generalization in which I and C are replaced by the arbitrary cubes I^X and C^X, respectively.

<u>THEOREM 2.11</u>. Let $S \in \underline{Z}$ and let X be an arbitrary set. If $f \in \mathrm{Sur}(S,I^X)$ and $\varepsilon > 0$, then there are a $g \in \mathrm{Sur}(S,C^X)$ and a family $\{h_x \in \mathrm{Sur}(I,I) : x \in X\}$ such that

i) $(\pi h_x)f = \rho^X g$; i.e. the following diagram commutes

$$\begin{array}{ccc} S & \xrightarrow{\quad f \quad} & I^X \\ {\scriptstyle g}\downarrow & & \downarrow{\scriptstyle \pi h_x} \\ C^X & \xrightarrow[\rho^X]{\qquad} & I^X \end{array}$$

(ii) $r \leq h_x(r) < r + \varepsilon$ for each $r \in I$ and each $x \in X$.

Proof. By 2.9 we find a surmorphism $g_x : S \to C$ and an $h_x \in \mathrm{Sur}(I,I)$ with $\rho g_x = h_x \, pr_x \, f$ such that ii) is satisfied for each $x \in X$. Then $(\pi h_x) f : S \to I^X$ is a surmorphism equalling $\rho^X g$ with $g(s) = (g_x(s))_{x \in X}$. Now $\rho^{-1}(\rho(c)) = \{c\}$ for each $c \in C$ which is neither a local minimum nor a local maximum in C. Thus the same is true for each point $c \in C^X$ each of whose coordinates is neither a local minimum nor a local maximum in C. But the set of all such points is dense in C^X, and since $\rho^X g$ is surjective, this set is in $g(S)$. Hence $C^X \subseteq g(S)$, so g is indeed surjective. $\square$

If $\mathrm{Sur}(S,\rho^X) : \mathrm{Sur}(S,C^X) \to \mathrm{Sur}(S,I^X)$ is the map defined by $\mathrm{Sur}(S,\rho^X)(f) = \rho^X f$, then Theorem 2.10 states that $\mathrm{Sur}(S,\rho^X)$ has dense image in $\mathrm{Sur}(S,I^X)$ when $\mathrm{Sur}(S,I^X)$ is equipped with the topology of uniform convergence relative to the metric d on I^X given by $d((i_x)_{x \in X},(j_x)_{x \in X}) = \sup\{|i_x - j_x| : x \in X\}$. A natural question is whether $\mathrm{Sur}(S,\rho^X)$ is actually surjective, i.e. does every $f \in \mathrm{Sur}(S,I^X)$ factor through ρ^X? The following example shows that this is not the case even when

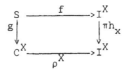

X is singleton.

EXAMPLE 2.12. Consider the graph $G = \{(x,\rho(x)) : x \in C\} \subseteq I \times I$ of the Cantor function $\rho : C \to I$. Then G is a semilattice under the multiplication induced from $I \times I$, and $K(G) = \{(x,\rho(x)) : x \in K(C)\}$. For each $k \in K(C)$, pick a sequence $\{y_n(k)\}_{n=1}^{\infty} \subseteq I$ which converges to $\rho(k)$ from below. Since $K(C)$ is countable, this can be accomplished so that $y_n(k) \notin Q$ and $y_n(k) = y_{n'}(k')$ implies $n = n'$ and $k = k'$ for all n and all $k,k' \in K(C)$. Thus there is a unique $x_n(k)$ with $\rho(x_n(k)) = y_n(k)$ for each n and each $k \in K(C)$. If $S = G \cup \{(k,y_n(k)) : k \in K(C), n = 1,2,\ldots\}$, then S is a compact zero dimensional space. If we give G and $\{(k,y_n(k)) : k \in K(C), n = 1,\ldots\}$ each the multiplication induced from $C \times I$, and if we define

$$(x,\rho(x))(k,y_n(k)) = (k,y_n(k))(x,\rho(x))$$
$$= \begin{cases} (k,\ y_n(k)) & \text{if } k < x \\ (x \wedge x_n(k),\rho(x \wedge x_n(k))) & \text{if } x < k, \end{cases}$$

then this multiplication makes S into a $\underline{Z}$-object. Moreover, $f : S \to I$ given by $f(x,y) = y$ is a surmorphism. However, there is no $g \in \mathrm{Sur}(S,C)$ with $f = \rho g$. Indeed, suppose $g \in \mathrm{Sur}(S,C)$, and let $k' \in K(C)$. Then $g^{-1}(\uparrow k')$ is an open closed subsemilattice of S, and so there is $s \in K(S)$ with $g^{-1}(\uparrow k') = \uparrow s$. Now $K(S) = \{(k,y_n(k)) : k \in K(C), n = 1,\ldots\}$ by construction, so $s = (k,y_n(k))$. Now, $\rho(k') \in Q$, but $f(k,y_n(k)) = y_n(k) \notin Q$ by choice of $y_n(k)$. Thus we cannot have $\rho g = f$ since $g(k,y_n(k)) = k'$. □

This example still leaves open the possibility that there is a large class of $\underline{Z}$-objects S for which $\mathrm{Sur}(S,\rho^X)$ is a surjection, and the following theorem demonstrates that this is indeed the case.

<u>THEOREM 2.13</u>. Let $S \in \underline{Z}$ and suppose S satisfies the condition

(DC) The dual semilattice $\hat{S}$ is a complete

semilattice (hence a complete lattice).
Then $Sur(S,\rho^X) : Sur(S,C^X) \to Sur(S,I^X)$ is surjective; i.e. for each surmorphism $f : S \to I^X$, there is a surmorphism $g : S \to C^X$ with $f = \rho^X g$. □

We now turn our attention to the problem of determining when $Sur(S,I)$ separates the points of S for a $\underline{Z}$-object S.

DEFINITION 2.14. An object $S \in \underline{Z}$ is called totally instable if $Sur(S,I)$ separates the points of S.

Obviously, every totally instable object is instable, but the product $S \times T$ of any non-degenerate stable object S and any instable object T shows that the converse is false. Moreover, as in the case $S = C$, it may occur that $Sur(S,C)$ separates the points of S while S is not totally instable.

The following definition is needed to characterize the separation of points by $Sur(S,C)$, respectively $Sur(S,I)$ in terms of duality.

DEFINITION 2.15. Let $S \in \underline{S}$. The set of all countable order dense chains $L \subseteq S$ containing 1 and having no minimum is denoted $c(S)$, and the subsemilattice of S generated by $\cup c(S)$ is denoted $A(S)$.

PROPOSITION 2.16. For a semilattice $S \in \underline{Z}$, the following are equivalent:

(1) $Sur(S,C)$ separates the points of S.

(2) For each injection $j : 3 \to S$, there is an $f \in Sur(S,C)$ with $fj : 3 \to C$ an injection, where $3 = \{0,\frac{1}{2},1\} \subseteq I$.

(3) For each surmorphism $\hat{j} : \hat{S} \to 3$ there is an $L \in c(\hat{S})$ with $\hat{j}(L) = 3$.

(4) Whenever $F \subsetneq G$ are proper filters on $\hat{S}$, there is an $L \in C(\hat{S})$ with $L \nsubseteq G$ and $L \cap (G\ F) \neq \emptyset$.

Furthermore, the following statements are equivalent and imply (1)-(4) above.

(I) S is totally instable.

(II) For each injection $j : 3 \to S$ there are $g \in Sur(S,C)$ and $e \in Sur(C,3)$ with $gj : 3 \to C$

an injection and $egj : 3 \to 3$ satisfying $egj(0) = 0$ and $egj(\frac{1}{2}) = egj(1) = 1$.

(III) For each surmorphism $\hat{j} : \hat{S} \to 3$ there are injections $\hat{g} : Q \to \hat{S}$ and $\hat{e} : 3 \to Q$ such that $\hat{j}\hat{g}$ is a surjection and $\hat{j}\hat{g}\hat{e} : 3 \to 3$ satisfies $\hat{j}\hat{g}\hat{e}(0) = \hat{j}\hat{g}\hat{e}(\frac{1}{2}) = \frac{1}{2}$.

(IV) Whenever $F \subsetneq G$ are proper filters on $\hat{S}$, there is an $L \in C(\hat{S})$ such that $L \subseteq G$ and $\mathrm{card}(L \cap (G \setminus F)) > 1$.

Note that condition (4) implies that there is an abundance of chains in S. In particular, if $\hat{S} = \cup c(\hat{S})$, then (4) is satisfied: Indeed, if $F \subsetneq G$ and proper filters on $\hat{S}$ and $x \in G \setminus F$ pick $L_x \in c(S)$ with $x \in L_x$. Now, let $y \leq x$ with $y \notin G$, and pick $L_y \in c(S)$ with $y \in L_y$. Then $L = [(L_y \cap S_y) \setminus \{y\}] \cup (L_x \cap \uparrow x)$ is a chain in $c(S)$ meeting $S \setminus G$ and $G \setminus F$. The converse of this statement is not true, as is demonstrated by

EXAMPLE 2.17. Let $Q_o = Q \cup \{0\}$, and let $S = \{(x,y) \in Q_o \times Q_o : x = 0 \text{ or } y = 0 \text{ but } x,y \neq 1\} \cup (1,1)$. Then 2.16 (4) is satisfied, whereas $\cup c(S) = S \setminus \{0\}$, which is not a subsemilattice. □

One might wonder at this point if there are any totally instable $\underline{Z}$-objects. If $S \in \underline{Z}$ has breadth 1, then S is a chain, and clearly S is not totally instable since, for each $k \in K(S)$ with $k \neq 0$, any $f \in \mathrm{Sur}(S,I)$ must identify k and k', where $k' = \sup Sk \setminus \{k\}$. The following is an example of a metric breadth 2 totally instable $\underline{Z}$-object.

EXAMPLE 2.18. Let $T \subseteq Q \times Q$ be the subsemilattice $T = Q' \times Q' \cup \{(1,1)\}$, where $Q' = Q \cap (0,1)$. If $F \subsetneq G$ are proper filters and $\mathrm{card}\, G \setminus F = 1$, then from the definition of a filter, we must have $G = F \cup \{\min G\}$. Let $(p,q) = \min G$. Since, for $a \in [p,1] \cap Q'$ and $b \in [q,1] \cap Q'$, $(a,q)(p,b) = (p,q)$, either $([p,1] \times q) \cap F = \emptyset$ or $(p \times [q,1]) \cap F = \emptyset$. Thus, condition IV of 2.16 is satisfied, so $S = \hat{T}$ is a totally instable $\underline{Z}$-

object, which is, in fact, isomorphic to
$C \times C/((\{0\} \times C) \cup (C \times \{0\}))$. □

For any compact $\underline{M}$-semilattice T, let $R_T = R_T(S)$ be the intersection of all kernel congruences of surmorphisms $S \to T$, where $S \in \underline{Z}$ (note that $R_T = S \times S$ if $\mathrm{Sur}(S,T) = \emptyset$ since the intersection over an empty collection of subsets of a set is the whose set). Let $q_S^T : S \to S/R_T$ be the natural quotient map in $\underline{M}$. The following is then immediate.

PROPOSITION 2.19. For any $S \in \underline{Z}$, $R_C(S) \subseteq R_I(S)$ (and $R_C(C) \neq R_I(C)$). Hence the following is a commutative diagram in $\underline{M}$.

$$\begin{array}{ccccc} & & S & & \\ & \swarrow^{q_C^S} & & \searrow^{q_I^S} & \\ S/R_C(S) & & \xrightarrow[p^S]{} & & S/R_I(S) \end{array}$$

with $q_C^S \in \underline{Z}$.
Every $f \in \mathrm{Sur}(S,C)$ factors uniquely through q_C^S, every $f \in \mathrm{Sur}(S,I)$ factors uniquely through q_I^S, and if every $f \in \mathrm{Sur}(S,C)$ factors through $q : S \to T$, then q factors through q_C^S (and similarly for I in place of C).
$\mathrm{Sur}(S/R_C(S),C)$ and $\mathrm{Sur}(S/R_I(S),I)$ separate the points of $S/R_C(S)$ and $S/R_I(S)$, respectively, and $S/R_C(S)$ and $S/R_I(S)$ are singleton if and only if S is stable.

COROLLARY 2.20. For $S \in \underline{Z}$, the inclusion $j_{\hat{S}} : A(\hat{S}) \to \hat{S}$ is dual to $q_C^S : S \to S/R_C(S)$. □

We also find that $S \in \underline{Z}$ is stable if and only if $A(\hat{S}) = \{1\}$, and $\mathrm{Sur}(S,C)$ separates the points of S if and only if $A(\hat{S}) = \hat{S}$. Moreover, if $S,T \in \underline{S}$ with $A(S) = A(T) = 1$, then $A(S \times T) = 1$, so that, if $\{S_i : i \in I\}$ is a family of $\underline{S}$-semilattices such that $A(S_i) = \{1\}$ for each $i \in I$, then $A(\coprod S_i) = \{1\}$. Since $\widehat{\coprod S_i} = \Pi \hat{S}_i$, we have the result that the product of stable $\underline{Z}$-objects is stable. The following example shows that this is not true for coproducts.

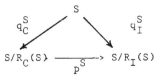

EXAMPLE 2.21. If X is a set, then $A(2^X) = \{1\}$ if and only if X is finite. Consequently, ${}^X 2$ (coproduct in $\underline{Z}$) is stable if and only if X is finite.

Proof. If X is finite, then, by the remarks above, $A(2^X) = \{1\}$ since $A(2) = \{1\}$. Thus, we only need show that, for X countably infinite, $A(2^X) \neq \{1\}$. Take $X = \hat{C} \simeq Q$ and define $f : C \to 2^X$ by $f(x)(\phi) = \phi(x)$. Since the characters separate the points of C, f is an injection, so 2^X contains a copy of C. Since $A(C) = C$, we have $A(2^X) \neq \{1\}$. □

Let $\underline{Z}_{stable}$ (respectively $\underline{Z}_{instable}$) denote the full subcategory of $\underline{Z}$ of all stable (respectively, instable) $\underline{Z}$-objects. Then we have the following summarizing result.

THEOREM 2.22. The category $\underline{Z}_{stable}$ is closed in $\underline{Z}$ under the formation of arbitrary products and under arbitrary quotients, but it fails to be closed under the formation of equalizers, pull-backs, projective limits (even strict ones), infinite products or injective limits. Likewise, the category $\underline{Z}_{instable}$ is closed in $\underline{Z}$ under the formation of arbitrary products and coproducts, and strict projective limits, but it fails to be closed under the formation of equalizers, pull-backs, or injective limits.

Section 3. Extremally disconnected compact semilattices.

It has been known for some time, that extremally disconnected (compact) groups are discrete. As a matter of fact, for this conclusion it suffices to know that all convergent sequences are eventually constant.

DEFINITION 3.1. A topological space X is called sequentially trivial, if for all sequences $(x_n)_{n=1,2,...}$ in X the relation $x = \lim x_n$ implies that only finitely many x_n differ from x. □

The following Lemma is rather well-known in this context.

LEMMA 3.2. Let X be a compact zero dimensional space such that the Boolean algebra of compact open sets of X is sigma complete (i.e. has countable sups), then X is sequentially trivial. In particular, any extremally disconnected space is sequentially trivial.

Proof (Indication). Let x_n be a convergent sequence which is not eventually constant. Assume that all x_n are different. Construct, by induction, a sequence of disjoint open closed neighborhoods U_n of x_n. Then $(\cup\{U_m \mid m > n\})^- \subseteq X\setminus(U_1 \cup \ldots \cup U_n)$. Now let $U = (\cup\{U_{2m-1} \mid m = 1,2,\ldots\})^-$. Then U is open closed by assumption on X. Now $x = \lim x_n \in U$ and $U_{2m} \cap U = \emptyset$ for all m. Hence $x \in (\cup U_{2m})^-$ is not contained in U. Contradiction. □

By contrast with extremal disconnectivity, sequential triviality is inherited by closed subspaces. We will show now that the finite objects are the only sequentially degenerate ones in $\underline{Z}$. We recall that for an element s in a semilattice S, we write $A(s)$ for the set of atoms in $\uparrow s$ (see III-2.8).

<u>THEOREM</u> <u>3.3</u>. Let S be a compact semilattice. Then the following statements are equivalent:

(1) S is extremally disconnected.

(2) S is sequentially trivial.

(3) S satisfies the ascending chain condition and $A(s)$ is finite for each $s \in S$.

(4) S is finite.

Proof. (4)=>(1) is trivial, (1)=>(2) is clear from 3.2, and (2)=>(3): Every ascending chain $x_1 \leq x_2 \leq \ldots$ converges by II-1.2 and hence is finite by (2).

We shall now show that all $A(s)$ are finite. By restricting our attention to $\uparrow s$ we may assume that $s = 0$. We show that $A(0)$ is finite. Clearly $A(0) \subseteq K(S)$ by II-3.3. Then all points of $A(0)$ are open in $A(0) \cup \{0\}$. Every element in S lies above an atom: If $x \in S$, then we set $x = x_1$, and if x_1 is not an atom, pick an element x_2 with $x_1 > x_2 > 0$. Continuing we create a sequence which must terminate after a finite number of steps since S is sequentially trivial. The last term is an atom below x. Now let $x \in (A(0) \cup \{0\})^- \setminus (A(0) \cup \{0\})$.

Then there is an $a \in A(0)$ with $a < x$ by what we just saw. Since $a \in K(S)$, then $\uparrow a$ is a neighborhood of x. This neighborhood must contain infinitely many elements of $A(0)$; but $\uparrow a \cap A(0) = \{a\}$. This contradiction shows that $A(0) \cup \{0\}$ is closed. Then $A(0) \cup \{0\}$ is finite or is the one point compactification of the discrete space $A(0)$. The latter, however, is not sequentially trivial. Hence $A(0)$ is finite.

(3)=>(4). Assume that S is infinite despite (3). Set $x_1 = 0$. Since $A(0)$ is finite, there is at least one $x_2 \in A(x_1)$ such that $\uparrow x_2$ is infinite. Since $A(x_2)$ is finite, there is an $x_3 \in A(x_2)$ with $\uparrow x_3$ infinite. Continue by induction and create a sequence $x_1 < x_2 < \ldots$. This is impossible since S satisfies the ascending chain condition. □

COROLLARY 3.4. Let S be a compact monoid which is a union of groups such that $E(S)$ is commutative. Then the following statements are equivalent:

(1) S is extremally disconnected.

(2) S is sequentially trivial.

(3) S is finite.

Proof. (2)=>(3): By Arhangelski's proof [A-1], every compact sequentially trivial group is finite. Hence all groups in S are finite. By 3.3, $E(S)$ is finite. Hence S is finite. The rest is trivial. □

Note that e.g. in this section we have shown that the space $\Gamma(E)$ of closed subsets of an extremally disconnected compact space E cannot itself be extremally disconnected or even sequentially trivial, since it is a topological semilattice under $\cup$.

HISTORICAL NOTES ON CHAPTER IV.

All of Chapter IV is new. The discussion of the cardinality invariants in Section 1 has been carried out completely for compact groups; as far as $d(G) = \log w(G)$ for compact groups is concerned, see Itzkowitz [I-3]. The results for semilattices agree with those for groups in parts and differ in others; in a sense they are not quite as pretty as in the group case, but they are just as conclusive in the presence of examples which we give. The theory in Section 2 is entirely new and is to be found in detail in [H-6]. Section 3 is new. The corresponding result for groups was most generally and directly proved by Arhangelski [A-1]; for compact groups this was perhaps proved the first time by Hofmann and Wright [H-10].

BIBLIOGRAPHY

A. 1. Arhangelski, A. V., Groups topologiques extrémalement discontinus, C. R. 265 (1967), 822-825.

2. Austin, C. W., Duality theorems for some commutative semigroups, Trans. A. M. S. 109 (1963), 245-256.

B. 1. Baker, J. W. and N. J. Rothman, Separating points by semicharacters in topological semigroups, Proc. A. M. S. 21 (1969), 235-239.

2. Baker, K. A., Inside free semilattices, Proc. Houston Conf. on Lattice Theory, 1973.

3. Balbes, R., Projective and injective distributive lattices, Pacific J. Math. 21 (1967), 405-420.

4. _________, Projective distributive lattices, Pacific J. Math. 33 (1970), 273-280.

5. _________, and A. Horn, Injective and projective Heyting algebras, Trans. A. M. S. 148 (1970), 549-560.

6. _________, and G. Grätzer, Injective and projective Stone algebras, Duke Math. J. 38 (1971), 339-347.

7. Banaschewski, B. and G. Bruns, Injective hulls in the category of distributive lattices, Journal R. und A. Math. 232 (1968), 102-109.

8. Birkhoff, G., Lattice Theory, A. M. S. Colloquium Pub., Providence R.I., (1967), xiii + 283 pp.

9. ___________, and O. Frink, Representations of lattices by sets, T.A.M.S. 64 (1948), 299-316.

10. Bowman, T., Analogue of Pontryagin character theory for topological semigroups, University of Florida, preprint (1973).

11. Bruns, G., Darstellungen und Erweiterung geordneten Mengen I, II, J. Reine Ang. Math. 209 (1962), 167-200, 210 (1962), 1-23.

12. Bulman-Fleming, S., On equationally compact semilattices, Algebra Universalis 2 (1972), 146-151.

D. 1. Davey, B. A., A note on representable posets, preprint.

2. Diener, K. H., Über zwei Birkhoff-Frinksche Struktursatze der allgemeinen Algebra, Arch. d. Math. 7 (1956), 337-345.

3. Dilworth, R. P. and P. Crawley, Decomposition theory for lattices without chain conditions, T.A.M.S. 96 (1960), 1-22.

G. 1. Gaskill, H. S., Classes of semilattices associated with an equational class of lattices, Can. J. Math. 25 (1973), 361-365.

2. Gilman, L. and M. Jerison, Rings of Continuous Functions, Van Nostrand, Princeton, N.J. (1960).

3. Gleason, A. M., Projective covers of topological spaces, Ill. J. of Math. 2 (1958), 482-489.

4. Grätzer, G., Lattice Theory, W. H. Freeman and Co., San Francisco, (1971), xv + 212 pp.

5. __________, Universal Algebra, D. van Nostrand, Toronto, (1968), xvi + 368 pp.

6. Green, C., A decomposition property for semilattices, Not. A.M.S. 15 (1968), 1040.

H. 1. Halmos, P. R., Injective and projective Boolean algebras, Proc. Sym. Pure Math. II, Lattice Theory, A.M.S. Publ. (1961), 114-122.

2. ____________, Boolean Algebras, Van Nostrand Math. Studies 1, Princeton, N.J., (1963), 147 pp.

3. Hofmann, K. H., Categories with convergence, exponential functors, and the cohomology of compact abelian groups, Math. Zeit. 104 (1968),

106-140.

4. Hofmann, K. H., The Duality of Compact Semigroups and C*-Bigebras, Lecture Notes in Math. 129, Springer-Verlag, Heidelberg (1970), xii + 142 pp.

5. _____________ and K. Keimel, A general character theory for partially ordered sets and lattices, Mem. A.M.S. 122 (1972).

6. _____________, M. Mislove, and A. Stralka, Dimension raising maps in topological algebra, Math. Zeit., to appear, 36 pp.

7. _____________ and P. S. Mostert, Elements of Compact Semigroups, Charles Merrill, Columbus, Ohio, (1966), xiii + 384 pp.

8. _____________ and A. Stralka, Mapping cylinders and compact monoids, Math. Annal. (1973), to appear.

9. _____________ and _________, Push-outs and strict projective limits of semilattices, Semigroup Forum 5 (1973), 243-262.

10. _____________ and F. B. Wright, Pointwise periodic groups, Fund. Math. 52 (1963) 103-122.

11. Horn, A. and N. Kimura, The category of semilattices, Algebra Universalis 1 (1971), 26-38.

I. 1. Isbell, J. R., Atomless parts of space, Math. Scand. 31 (1972), 5-32.

2. ____________, General functorial semantics I, Amer. J. Math. 94 (1972), 535-596.

3. Itzkowitz, G. L., On the density character of compact groups, Fund. Math. 75 (1973), 201-203.

K. 1. Katětov, M., Remarks on Boolean algebras, Coll. Math. 2 (1951), 229-235.

2. Kuratowski, K., Topology, Vol. I, Academic Press, New York (1966), 560 + xx pp.

L. 1. Lawson, J. D., Intrinsic topologies in topological lattices and semilattices, Pacific J. Math. 44 (1973), 593-602.

2. Lawson, J. D., Intrinsic lattice and semilattice topologies, preprint.

M. 1. Maeda, F., Kontinuierliche Geometrien, Springer-Verlag, Heidelberg, (1958), x + 244 pp.

2. Martinez, J., The structure of archimedean lattices, Univ. of Florida, preprint (1973).

3. ________, Topological coordinatizations and dualities of Brouwer lattices, Univ. of Florida, preprint (1973).

4. Mena, R. A., Ideal completions and lattices of ideals, Dissertation, Univ. of Houston, (1973).

5. Mitchell, B., Category Theory, Academic Press, New York, (1965), xi + 273 pp.

N. 1. Nachbin, L., On characterizations of the lattice of all ideals of a ring, Fund. Math. 6 (1949), 137-142.

2. Numakura, K., Theorems on compact totally disconnected semigroups and lattices, Proc. A.M.S. 8 (1957), 623-626.

P. 1. Papert Strauss, Dona, Topological lattices, Proc. Lon. Math. Soc. (3) 18 (1968), 217-230.

2. Priestley, H. A., Ordered topological spaces and the representation of distributive lattices, Proc. Lon. Math. Soc. (3) 24 (1972), 507-530.

3. ________, Representations of distributive lattices by means of ordered Stone spaces, Bull. Lon. Math. Soc. 2 (1970), 186-190.

R. 1. Raney, G. N., A subdirect-union representation for completely distributive complete lattices, Proc. A.M.S. 4 (1953), 518-522.

2. _________, Completely distributive complete lattices, Proc. A.M.S. 3 (1952), 677-680.

3. Roeder, D. W., Category theory applied to Pontryagin duality, Pac. J. Math,, to appear.

4. Rhodes, J., Decomposition semilattices with applications to topological lattices, Pac. J. Math. 44 (1973), 299-307.

S. 1. Schein, B. M., On the definition of distributive semilattices, Alg. Univ. 2 (1972), 1-2.

2. Schmidt, J., Universal and internal properties of partially ordered sets, Journal R. und A. Math. 253 (1972), 28-42.

3. _________, Universal and internal properties of some completions of k-join-semilattices and k-join distributive partially ordered sets, Journal R. und A. Math. 255 (1972), 8-22.

4. Schneperman, L. B., On the theory of characters of locally bicompact topological semigroups, Math. Sbor. 77 (1968), 508-532, (Math. USSR-Sb. 6 (1968), 471-492.

5. Scott, Dana, Continuous lattices, Toposes, Algebraic Geometry, and Logic, Springer Lecture Notes 274, 97-136.

6. Sikorski, R., Boolean Algebras, 2nd ed., Springer Verlag, Heidelberg, (1964).

7. Stralka, A., The lattice of ideals of a compact semilattice, Proc. A.M.S. 33 (1972), 175-180.

8. _________, Distributive topological lattices, to appear.

T. 1. Tarski, A., Sur les classes closes par rapport à certaines operations élémentaires, Fund. Math. 16 (1929), 181-305.

V. 1. Varlet, J., On distributive residerated groupoids, Semigroup Forum 6 (1973), 80-85.

2. _________, Modularity and distributivity in partially ordered groupoids, Bull. Soc. Roy. Sci. Liège 38 (1969), 639-648.

W. 1. Weil, A., Basic Number Theory, Springer-Verlag, Heidelberg, 1967, xi + 294 pp.

NOTATION

INDEX

Vol. 215: P. Antonelli, D. Burghelea and P. J. Kahn, The Concordance-Homotopy Groups of Geometric Automorphism Groups. X, 140 pages. 1971. DM 16,-

Vol. 216: H. Maaß, Siegel's Modular Forms and Dirichlet Series. VII, 328 pages. 1971. DM 20,-

Vol. 217: T. J. Jech, Lectures in Set Theory with Particular Emphasis on the Method of Forcing. V, 137 pages. 1971. DM 16,-

Vol. 218: C. P. Schnorr, Zufälligkeit und Wahrscheinlichkeit. IV, 212 Seiten. 1971. DM 20,-

Vol. 219: N. L. Alling and N. Greenleaf, Foundations of the Theory of Klein Surfaces. IX, 117 pages. 1971. DM 16,-

Vol. 220: W. A. Coppel, Disconjugacy. V, 148 pages. 1971. DM 16,-

Vol. 221: P. Gabriel und F. Ulmer, Lokal präsentierbare Kategorien. V, 200 Seiten. 1971. DM 18,-

Vol. 222: C. Meghea, Compactification des Espaces Harmoniques. III, 108 pages. 1971. DM 16,-

Vol. 223: U. Felgner, Models of ZF-Set Theory. VI, 173 pages. 1971. DM 16,-

Vol. 224: Revètements Etales et Groupe Fondamental. (SGA 1). Dirigé par A. Grothendieck XXII, 447 pages. 1971. DM 30,-

Vol. 225: Théorie des Intersections et Théorème de Riemann-Roch. (SGA 6). Dirigé par P. Berthelot, A. Grothendieck et L. Illusie. XII, 700 pages. 1971. DM 40,-

Vol. 226: Seminar on Potential Theory, II. Edited by H. Bauer. IV, 170 pages. 1971. DM 18,-

Vol. 227: H. L. Montgomery, Topics in Multiplicative Number Theory. IX, 178 pages. 1971. DM 18,-

Vol. 228: Conference on Applications of Numerical Analysis. Edited by J. Ll. Morris. X, 358 pages. 1971. DM 26,-

Vol. 229: J. Väisälä, Lectures on n-Dimensional Quasiconformal Mappings. XIV, 144 pages. 1971. DM 16,-

Vol. 230: L. Waelbroeck, Topological Vector Spaces and Algebras. VII, 158 pages. 1971. DM 16,-

Vol. 231: H. Reiter, L^1-Algebras and Segal Algebras. XI, 113 pages. 1971. DM 16,-

Vol. 232: T. H. Ganelius, Tauberian Remainder Theorems. VI, 75 pages. 1971. DM 16,-

Vol. 233: C. P. Tsokos and W. J. Padgett. Random Integral Equations with Applications to stochastic Systems. VII, 174 pages. 1971. DM 18,-

Vol. 234: A. Andreotti and W. Stoll. Analytic and Algebraic Dependence of Meromorphic Functions. III, 390 pages. 1971. DM 26,-

Vol. 235: Global Differentiable Dynamics. Edited by O. Hájek, A. J. Lohwater, and R. McCann. X, 140 pages. 1971. DM 16,-

Vol. 236: M. Barr, P. A. Grillet, and D. H. van Osdol. Exact Categories and Categories of Sheaves. VII, 239 pages. 1971. DM 20,-

Vol. 237: B. Stenström, Rings and Modules of Quotients. VII, 136 pages. 1971. DM 16,-

Vol. 238: Der kanonische Modul eines Cohen-Macaulay-Rings. Herausgegeben von Jürgen Herzog und Ernst Kunz. VI, 103 Seiten. 1971. DM 16,-

Vol. 239: L. Illusie, Complexe Cotangent et Déformations I. XV, 355 pages. 1971. DM 26,-

Vol. 240: A. Kerber, Representations of Permutation Groups I. VII, 92 pages. 1971. DM 18,-

Vol. 241: S. Kaneyuki, Homogeneous Bounded Domains and Siegel Domains. V, 89 pages. 1971. DM 16,-

Vol. 242: R. R. Coifman et G. Weiss, Analyse Harmonique Non-Commutative sur Certains Espaces. V, 160 pages. 1971. DM 16,-

Vol. 243: Japan-United States Seminar on Ordinary Differential and Functional Equations. Edited by M. Urabe. VIII, 332 pages. 1971. DM 26,-

Vol. 244: Séminaire Bourbaki - vol. 1970/71. Exposés 382-399. 356 pages. 1971. DM 26,-

Vol. 245: D. E. Cohen, Groups of Cohomological Dimension One. V, pages. 1972. DM 16,-

Vol. 246: Lectures on Rings and Modules. Tulane University Ring and Operator Theory Year, 1970-1971. Volume I. X, 661 pages. 1972. DM 40,-

Vol. 247: Lectures on Operator Algebras. Tulane University Ring and Operator Theory Year, 1970-1971. Volume II. XI, 786 pages. 1972. DM 40,-

Vol. 248: Lectures on the Applications of Sheaves to Ring Theory. Tulane University Ring and Operator Theory Year, 1970-1971. Volume III. VIII, 315 pages. 1971. DM 26,-

Vol. 249: Symposium on Algebraic Topology. Edited by P. J. Hilton. VII, 111 pages. 1971. DM 16,-

Vol. 250: B. Jónsson, Topics in Universal Algebra. VI, 220 pages. 1972. DM 20,-

Vol. 251: The Theory of Arithmetic Functions. Edited by A. A. Gioia and D. L. Goldsmith VI, 287 pages. 1972. DM 24,-

Vol. 252: D. A. Stone, Stratified Polyhedra. IX, 193 pages. 1972. DM 18,-

Vol. 253: V. Komkov, Optimal Control Theory for the Damping of Vibrations of Simple Elastic Systems. V, 240 pages. 1972. DM 20,-

Vol. 254: C. U. Jensen, Les Foncteurs Dérivés de $\varprojlim$ et leurs Applications en Théorie des Modules. V, 103 pages. 1972. DM 16,-

Vol. 255: Conference in Mathematical Logic - London '70. Edited by W. Hodges. VIII, 351 pages. 1972. DM 26,-

Vol. 256: C. A. Berenstein and M. A. Dostal, Analytically Uniform Spaces and their Applications to Convolution Equations. VII, 130 pages. 1972. DM 16,-

Vol. 257: R. B. Holmes, A Course on Optimization and Best Approximation. VIII, 233 pages. 1972. DM 20,-

Vol. 258: Séminaire de Probabilités VI. Edited by P. A. Meyer. VI, 253 pages. 1972. DM 22,-

Vol. 259: N. Moulis, Structures de Fredholm sur les Variétés Hilbertiennes. V, 123 pages. 1972. DM 16,-

Vol. 260: R. Godement and H. Jacquet, Zeta Functions of Simple Algebras. IX, 188 pages. 1972. DM 18,-

Vol. 261: A. Guichardet, Symmetric Hilbert Spaces and Related Topics. V, 197 pages. 1972. DM 18,-

Vol. 262: H. G. Zimmer, Computational Problems, Methods, and Results in Algebraic Number Theory. V, 103 pages. 1972. DM 16,-

Vol. 263: T. Parthasarathy, Selection Theorems and their Applications. VII, 101 pages. 1972. DM 16,-

Vol. 264: W. Messing, The Crystals Associated to Barsotti-Tate Groups: With Applications to Abelian Schemes. III, 190 pages. 1972. DM 18,-

Vol. 265: N. Saavedra Rivano, Catégories Tannakiennes. II, 418 pages. 1972. DM 26,-

Vol. 266: Conference on Harmonic Analysis. Edited by D. Gulick and R. L. Lipsman. VI, 323 pages. 1972. DM 24,-

Vol. 267: Numerische Lösung nichtlinearer partieller Differential- und Integro-Differentialgleichungen. Herausgegeben von R. Ansorge und W. Törnig, VI, 339 Seiten. 1972. DM 26,-

Vol. 268: C. G. Simader, On Dirichlet's Boundary Value Problem. IV, 238 pages. 1972. DM 20,-

Vol. 269: Théorie des Topos et Cohomologie Etale des Schémas. (SGA 4). Dirigé par M. Artin, A. Grothendieck et J. L. Verdier. XIX, 525 pages. 1972. DM 50,-

Vol. 270: Théorie des Topos et Cohomologie Etale des Schémas. Tome 2. (SGA 4). Dirigé par M. Artin, A. Grothendieck et J. L. Verdier. V, 418 pages. 1972. DM 50,-

Vol. 271: J. P. May, The Geometry of Iterated Loop Spaces. IX, 175 pages. 1972. DM 18,-

Vol. 272: K. R. Parthasarathy and K. Schmidt, Positive Definite Kernels, Continuous Tensor Products, and Central Limit Theorems of Probability Theory. VI, 107 pages. 1972. DM 16,-

Vol. 273: U. Seip, Kompakt erzeugte Vektorräume und Analysis. IX, 119 Seiten. 1972. DM 16,-

Vol. 274: Toposes, Algebraic Geometry and Logic. Edited by. F. W. Lawvere. VI, 189 pages. 1972. DM 18,-

Vol. 275: Séminaire Pierre Lelong (Analyse) Année 1970-1971. VI, 181 pages. 1972. DM 18,-

Vol. 276: A. Borel, Représentations de Groupes Localement Compacts. V, 98 pages. 1972. DM 16,-

Vol. 277: Séminaire Banach. Edité par C. Houzel. VII, 229 pages. 1972. DM 20,-

Vol. 278: H. Jacquet, Automorphic Forms on GL(2). Part II. XIII, 142 pages. 1972. DM 16,–

Vol. 279: R. Bott, S. Gitler and I. M. James, Lectures on Algebraic and Differential Topology. V, 174 pages. 1972. DM 18,–

Vol. 280: Conference on the Theory of Ordinary and Partial Differential Equations. Edited by W. N. Everitt and B. D. Sleeman. XV, 367 pages. 1972. DM 26,–

Vol. 281: Coherence in Categories. Edited by S. Mac Lane. VII, 235 pages. 1972. DM 20,–

Vol. 282: W. Klingenberg und P. Flaschel, Riemannsche Hilbertmannigfaltigkeiten. Periodische Geodätische. VII, 211 Seiten. 1972. DM 20,–

Vol. 283: L. Illusie, Complexe Cotangent et Déformations II. VII, 304 pages. 1972. DM 24,–

Vol. 284: P. A. Meyer, Martingales and Stochastic Integrals I. VI, 89 pages. 1972. DM 16,–

Vol. 285: P. de la Harpe, Classical Banach-Lie Algebras and Banach-Lie Groups of Operators in Hilbert Space. III, 160 pages. 1972. DM 16,–

Vol. 286: S. Murakami, On Automorphisms of Siegel Domains. V, 95 pages. 1972. DM 16,–

Vol. 287: Hyperfunctions and Pseudo-Differential Equations. Edited by H. Komatsu. VII, 529 pages. 1973. DM 36,–

Vol. 288: Groupes de Monodromie en Géométrie Algébrique. (SGA 7 I). Dirigé par A. Grothendieck. IX, 523 pages. 1972. DM 50,–

Vol. 289: B. Fuglede, Finely Harmonic Functions. III, 188. 1972. DM 18,–

Vol. 290: D. B. Zagier, Equivariant Pontrjagin Classes and Applications to Orbit Spaces. IX, 130 pages. 1972. DM 16,–

Vol. 291: P. Orlik, Seifert Manifolds. VIII, 155 pages. 1972. DM 16,–

Vol. 292: W. D. Wallis, A. P. Street and J. S. Wallis, Combinatorics: Room Squares, Sum-Free Sets, Hadamard Matrices. V, 508 pages. 1972. DM 50,–

Vol. 293: R. A. DeVore, The Approximation of Continuous Functions by Positive Linear Operators. VIII, 289 pages. 1972. DM 24,–

Vol. 294: Stability of Stochastic Dynamical Systems. Edited by R. F. Curtain. IX, 332 pages. 1972. DM 26,–

Vol. 295: C. Dellacherie, Ensembles Analytiques, Capacités, Mesures de Hausdorff. XII, 123 pages. 1972. DM 16,–

Vol. 296: Probability and Information Theory II. Edited by M. Behara, K. Krickeberg and J. Wolfowitz. V, 223 pages. 1973. DM 20,–

Vol. 297: J. Garnett, Analytic Capacity and Measure. IV, 138 pages. 1972. DM 16,–

Vol. 298: Proceedings of the Second Conference on Compact Transformation Groups. Part 1. XIII, 453 pages. 1972. DM 32,–

Vol. 299: Proceedings of the Second Conference on Compact Transformation Groups. Part 2. XIV, 327 pages. 1972. DM 26,–

Vol. 300: P. Eymard, Moyennes Invariantes et Représentations Unitaires. II. 113 pages. 1972. DM 16,–

Vol. 301: F. Pittnauer, Vorlesungen über asymptotische Reihen. VI, 186 Seiten. 1972. DM 18,–

Vol. 302: M. Demazure, Lectures on p-Divisible Groups. V, 98 pages. 1972. DM 16,–

Vol. 303: Graph Theory and Applications. Edited by Y. Alavi, D. R. Lick and A. T. White. IX, 329 pages. 1972. DM 26,–

Vol. 304: A. K. Bousfield and D. M. Kan, Homotopy Limits, Completions and Localizations. V, 348 pages. 1972. DM 26,–

Vol. 305: Théorie des Topos et Cohomologie Etale des Schémas. Tome 3. (SGA 4). Dirigé par M. Artin, A. Grothendieck et J. L. Verdier. VI, 640 pages. 1973. DM 50,–

Vol. 306: H. Luckhardt, Extensional Gödel Functional Interpretation. VI, 161 pages. 1973. DM 18,–

Vol. 307: J. L. Bretagnolle, S. D. Chatterji et P.-A. Meyer, Ecole d'été de Probabilités: Processus Stochastiques. VI, 198 pages. 1973. DM 20,–

Vol. 308: D. Knutson, λ-Rings and the Representation Theory of the Symmetric Group. IV, 203 pages. 1973. DM 20,–

Vol. 309: D. H. Sattinger, Topics in Stability and Bifurcation Theory. VI, 190 pages. 1973. DM 18,–

Vol. 310: B. Iversen, Generic Local Structure of the Morphisms in Commutative Algebra. IV, 108 pages. 1973. DM 16,–

Vol. 311: Conference on Commutative Algebra. Edited by J. W. Brewer and E. A. Rutter. VII, 251 pages. 1973. DM 22,–

Vol. 312: Symposium on Ordinary Differential Equations. Edited by W. A. Harris, Jr. and Y. Sibuya. VIII, 204 pages. 1973. DM 22,–

Vol. 313: K. Jörgens and J. Weidmann, Spectral Properties of Hamiltonian Operators. III, 140 pages. 1973. DM 16,–

Vol. 314: M. Deuring, Lectures on the Theory of Algebraic Functions of One Variable. VI, 151 pages. 1973. DM 16,–

Vol. 315: K. Bichteler, Integration Theory (with Special Attention to Vector Measures). VI, 357 pages. 1973. DM 26,–

Vol. 316: Symposium on Non-Well-Posed Problems and Logarithmic Convexity. Edited by R. J. Knops. V, 176 pages. 1973. DM 18,–

Vol. 317: Séminaire Bourbaki – vol. 1971/72. Exposés 400–417. IV, 361 pages. 1973. DM 26,–

Vol. 318: Recent Advances in Topological Dynamics. Edited by A. Beck, VIII, 285 pages. 1973. DM 24,–

Vol. 319: Conference on Group Theory. Edited by R. W. Gatterdam and K. W. Weston. V, 188 pages. 1973. DM 18,–

Vol. 320: Modular Functions of One Variable I. Edited by W. Kuyk. V, 195 pages. 1973. DM 18,–

Vol. 321: Séminaire de Probabilités VII. Edité par P. A. Meyer. VI, 322 pages. 1973. DM 26,–

Vol. 322: Nonlinear Problems in the Physical Sciences and Biology. Edited by I. Stakgold, D. D. Joseph and D. H. Sattinger. VIII, 357 pages. 1973. DM 26,–

Vol. 323: J. L. Lions, Perturbations Singulières dans les Problèmes aux Limites et en Contrôle Optimal. XII, 645 pages. 1973. DM 42,–

Vol. 324: K. Kreith, Oscillation Theory. VI, 109 pages. 1973. DM 16,–

Vol. 325: Ch.-Ch. Chou, La Transformation de Fourier Complexe et L'Equation de Convolution. IX, 137 pages. 1973. DM 16,–

Vol. 326: A. Robert, Elliptic Curves. VIII, 264 pages. 1973. DM 22,–

Vol. 327: E. Matlis, 1-Dimensional Cohen-Macaulay Rings. XII, 157 pages. 1973. DM 18,–

Vol. 328: J. R. Büchi and D. Siefkes, The Monadic Second Order Theory of All Countable Ordinals. VI, 217 pages. 1973. DM 20,–

Vol. 329: W. Trebels, Multipliers for (C, α)-Bounded Fourier Expansions in Banach Spaces and Approximation Theory. VII, 103 pages. 1973. DM 16,–

Vol. 330: Proceedings of the Second Japan-USSR Symposium or Probability Theory. Edited by G. Maruyama and Yu. V. Prokhorov VI, 550 pages. 1973. DM 36,–

Vol. 331: Summer School on Topological Vector Spaces. Edited by L. Waelbroeck. VI, 226 pages. 1973. DM 20,–

Vol. 332: Séminaire Pierre Lelong (Analyse) Année 1971-1972. V, 13 pages. 1973. DM 16,–

Vol. 333: Numerische, insbesondere approximationstheoretische Be handlung von Funktionalgleichungen. Herausgegeben von R. An sorge und W. Törnig. VI, 296 Seiten. 1973. DM 24,–

Vol. 334: F. Schweiger, The Metrical Theory of Jacobi-Perron Algo rithm. V, 111 pages. 1973. DM 16,–

Vol. 335: H. Huck, R. Roitzsch, U. Simon, W. Vortisch, R. Walden, Wegner und W. Wendland, Beweismethoden der Differentialgeome trie im Großen. IX, 159 Seiten. 1973. DM 18,–

Vol. 336: L'Analyse Harmonique dans le Domaine Complexe. Edi par E. J. Akutowicz. VIII, 169 pages. 1973. DM 18,–

Vol. 337: Cambridge Summer School in Mathematical Logic. Edit by A. R. D. Mathias and H. Rogers. IX, 660 pages. 1973. DM 42

Vol: 338: J. Lindenstrauss and L. Tzafriri, Classical Banach Space IX, 243 pages. 1973. DM 22,–

Vol. 339: G. Kempf, F. Knudsen, D. Mumford and B. Saint-Don Toroidal Embeddings I. VIII, 209 pages. 1973. DM 20,–

Vol. 340: Groupes de Monodromie en Géométrie Algébriqu (SGA 7 II). Par P. Deligne et N. Katz. X, 438 pages. 1973. DM 40

Vol. 341: Algebraic K-Theory I, Higher K-Theories. Edited H. Bass. XV, 335 pages. 1973. DM 26,–

Vol. 342: Algebraic K-Theory II, "Classical" Algebraic K-Theo and Connections with Arithmetic. Edited by H. Bass. XV, 5 pages. 1973. DM 36,–